Benno Müller-Hill
The *lac* Operon

Benno Müller-Hill

The *lac* Operon

A Short History of a Genetic Paradigm

Walter de Gruyter · Berlin · New York 1996

Professor Benno Müller-Hill
Institut für Genetik an der Universität zu Köln
Weyertal 121, D-50931 Köln, Germany

With 20 figures and 2 tables.

Cover Illustration
Lac repressor-16 bp operator cocrystal in stabilizing buffer (left) and instabilizing buffer after adding IPTG (right).
(see: Pace et al. (1990) Proceedings of the National Academy of Sciences USA *87*, 1870 – 1873; courtesy of Ponzy Lu)

∞ Printed on acid-free paper which falls within the guidelines of the ANSI to ensure permanence and durability.

Library of Congress Cataloging-in-Publication Data

Müller-Hill, Benno, 1933 –
    The lac Operon : a short history of a genetic paradigm /
Benno Müller-Hill
        p.   cm.
    Includes index.
    ISBN 3-11-014830-7 (pb. ; alk. paper)
    1. Lac operon – Research – History.   I. Title.
QH434.M85   1996
589.9′5 – dc20   96-20965
                                                         CIP

Die Deutsche Bibliothek – CIP Einheitsaufnahme

Müller-Hill, Benno
The lac operon : a short history of a genetic paradigm / Benno
Müller-Hill. – Berlin ; New York : de Gruyter, 1996
    ISBN 3-11-014830-7

Typesetting: Knipp Medien und Kommunikation, Dortmund. – Printing: Ratzlow-Druck, Berlin. – Binding: Mikolai GmbH, Berlin.
Cover Design: Hansbernd Lindemann, Berlin.

To
Jim and Wally
who taught me science

Contents

Part 3: The *lac* Operon, a Paradigm of Beauty and Efficiency . .   129

Introduction

This book presents a short history of the *lac* system of *Escherichia coli* (*E. coli*). This is unusual. Molecular biology has no history for the young scientist. What happened ten years ago seems prehistoric and thus of no interest. Recently I asked a student of the "Max Delbrück Laboratory" during his doctor's exam to describe the experiment which made Max famous. The student who had produced an excellent thesis did not know. He had never heard of Luria. "Phage" was all he could say. And indeed, the Luria-Delbrück experiment is no longer described in many current textbooks.

This gives the young scientist the illusion that only the new exists and that everything new is true. Thus, for the student molecular biology has two faces. In the textbooks, almost everything is solved and clear. Most claims are so self-evident that no proofs are given. Old, classical experiments disappear. They are self-evident. Old errors in interpretation are not mentioned. Who cares? There is only one view, and this is the correct, modern view. Since molecular biology is virtually devoid of mathematical theory, multicoloured diagrams are the condensed essence. Thus, the student has to memorize multicoloured pictures as the essence of molecular biology. And I dare say they are sometimes misleading.

The other face of molecular biology is seen at scientific conferences or read in recent issues of *Nature*, *Science* or *Cell*. There, material is presented which has not yet been included in textbooks – and which most likely never will be. The mass of papers is growing exponentially. Textbook authors have to select the papers they will present. They will have to believe the abstracts, they have no time to carefully study the results. So, errors and mistakes enter the textbooks, veiled as truth. Their number will increase with time. Will there be total confusion at the end? Will molecular biology become a kind of art or advertising, where anything goes?

One has to grow old to understand the functioning of science. Then one may remember the papers which a long time ago aroused great interest and one may compare them to the present-day textbooks. It is virtually impossible to do this with all of molecular biology. One has to concentrate on the small field one knows well. I will present here the lactose operon of *E. coli*, the system I know best. I do not pretend that this is *the* history of the *lac* system. History becomes smooth, too smooth in writing. Right starts are forgotten, wrong expeditions into the desert disappear without a trace. Only a few papers will be cited. The reader may wish to consult a collection of the selected scientific papers of Jacques Monod (1) and the commentaries of some of his collaborators (2). Another view is presented in the autobiography of François Jacob (3). In 1970, at the height of the classical era, a volume was dedicated solely to the *lac* operon (4). In 1978, at the beginning of the new era, about half of another volume (5) was dedicated to the *lac* operon, the rest dealt with other bacterial systems. Finally, the reader is alerted to the excellent book by Mark Ptashne on the competing λ system (6), a book I admire but did not want to copy. The reader will find that my description differs from the presentation in the textbooks. This clash is intended.

The peace and quiet of a sabbatical made it possible to concentrate on the writing of this book. I could not have written it without the support of the *Deutsche Forschungsgemeinschaft*, the *Ministerium für Forschung und Technologie* and the *Fonds der Chemie* who financed my research on this system for more than two decades. Many people read parts or all of the manuscript, typed meticulously by Elisabeth Stratmann and Anke Wagemann, and sent me their comments. I particularly thank Andrew Barker, Georges Cohen, Renate Dildrop, Walter Gilbert, Melvin Green, Jonathan Howard, George Klein, Peter Kolkhof, Howard Rickenberg, Maxim Schwartz and Agnes Ullmann. Any remaining errors are mine.

References

(1) Selected papers in molecular biology by Jacques Monod. Ed. by A. Lwoff & A. Ullmann, Academic Press, New York, San Francisco, London, 1978.

(2) Origins of molecular biology. A tribute to Jacques Monod. Ed. by A. Lwoff & A. Ullmann. Academic Press, New York, San Francisco. London, 1979.

(3) Jacob, F.: La statue intérieure. Éditions Odile Jacob, Paris 1987. The statue within: an autobiography. Basic Books, New York, 1988.

(4) The lactose operon. Ed. by J.R. Beckwith & D. Zipser. Cold Spring Harbor Laboratory, Cold Spring Harbor, New York, 1970.

(5) The operon. Ed. by J.H. Miller & W.S. Reznikoff. Cold Spring Harbor Laboratory, Cold Spring Harbor, N Y, 1978.

(6) Ptashne, M.: A genetic switch. Gene control and phage λ. Cell Press and Blackwell Scientific Publications. Cambridge, MA and Palo Alto, CA, 1986.

Part 1

A Short History of the *lac* System from its Beginning to 1978

1.1 From Noah to Pasteur: Adaptation in Yeast

The story begins four thousand years ago. Noah, having just escaped the deluge, planted a vineyard and produced wine (1). The sugar of grapes is turned into alcohol by omnipresent yeast. Later, yeast was isolated and used to make delicious bread. When the Jewish slaves fled from Egypt, they had no time to put yeast in their dough (2). When the Jews celebrate and commemorate the occasion on the Passover, they drink wine and eat unleavened bread (which is made without yeast). When Jesus celebrated the Passover with his students for the last time, he pointed to the bread and to the wine and said: "This is my body … this is my blood" (3). In the Latin Christian service "this is my body" became "*hoc est corpus*". In the eighteenth century, at Dutch fairs and market places, the tricksters called their tricks "*hocus pocus*".

What happens when the juice of grapes turns into wine? What happens in the bread? Antony van Leeuwenhoek (1632-1723) was the first to look at yeast with a microscope he had constructed (Leeuwenhoek, 1680, quoted in 4). He and the biologists believed that yeast was alive. Some chemists later doubted that. In fact, Justus von Liebig (1803-1873), professor at Giessen University, who studied alcohol fermentation, stated in many papers that fermentation was a chemical process, not coupled to living matter. He presumed that the yeast broke down into non-living pieces which carried out the breakdown of glucose and smashed it into ethanol. Liebig kept his opinion until his death, as evident from his last paper on the subject (5). He had already been attacked most convincingly in 1861 by Louis Pasteur (1822-1895), who had presented solid evidence that fermentation is intimately coupled to the living state of the yeast cells (4). Pasteur was right with his experiments, but Liebig was right with his intuition that fermentation was simple chemistry.

Two years after Pasteur's death, the German chemist Eduard Buchner (1860-1917), then professor at Tübingen University, demonstrated that a yeast extract, which shows no sign of living cells, is able to break down glucose to ethanol (6). How does this breakdown of glucose to ethanol occur? The molecule responsible for this process was called "zymase". But what happens if one uses, instead of glucose, the related sugar galactose? In 1900 Frédéric Dienert, working at the Pasteur Institute, showed that yeast grown on glucose breaks down glucose but not galactose, but yeast grown on galactose breaks down galactose or glucose (7). If yeast is grown in the presence of glucose *and* galactose, then the glucose is broken down first. It takes time until the yeast adapts, "accustoms", to galactose. Die-

[handwritten margin note: Pasteur explains fermentation is done by living yeast]

nert got similar results with other sugars. He also found yeast strains which were unable to break down galactose. Dienert discussed two possibilities to explain his observations: 1. there is just one fundamental molecule ("zymase") which is suitably modified to break down galactose; or 2. there is *de novo* synthesis of another zymase which now breaks down galactose (7). These experiments were repeated by the Nobel Prize winners Hans von Euler (1873-1964) and Richard Willstätter (1872-1942). There was no doubt about their correctness. But the emphasis shifted from regulatory aspects to the identification of the individual enzymes involved in these reactions.

References

(1) Genesis 9, 20.

(2) Exodus 12, 34.

(3) Matthew 26, 26-28.

(4) Pasteur, L.: Mémoire sur la fermentation alcoholique. Ann.de Chem.et de Phys. III Serie **58**, 323-426, 1861.

(5) Liebig, J.v.: Ueber die Gaehrung und die Quelle der Muskelkraft. Liebigs Annalen der Chemie **153**, 1-47, 1870.

(6) Buchner, E.: Alkoholische Gährung ohne Hefezellen. Berichte der Deutschen Chem.Ges. **30**, 117-124, 1897.

(7) Dienert, F.: Sur la fermentation du galactose et sur l'accoutumance des levures à ce sucre. Ann.Inst.Pasteur **14**, 139-189, 1900.

1.2 Adaptation in *Bacillus subtilis* and *Escherichia coli*

The *lac* story starts in the beginning of the winter of 1940. Paris was occupied by the German armed forces. Jacques Monod (1910-1976), then a young man of thirty years, asked André Lwoff (1902-1994) at the Institut Pasteur for advice. Four years earlier, he had returned from the USA where he had spent a year in the laboratory of Thomas Hunt Morgan working on genetics. *Drosophila* research seemed exhausted and had not become his subject. Now at the Sorbonne he had almost finished his thesis, dealing with bacterial growth. At that time, bacteria were generally thought not to have a nucleus and thus no genes. He had discovered a strange phenomenon. If he provided the bacteria with a mixture of two carbon sources, they sometimes used first one, then the other. What could it mean? "Adaptive enzymes", said Lwoff. "What are adaptive enzymes?" asked Jacques Monod (1,2).

André Lwoff may have told him that microorganisms such as bacteria and yeast adapt their metabolism, and thus their enzymes to the carbon source to which they are exposed. Many people had already worked on adaptation. A classical paper had been published in 1900 by Frédéric Dienert (3) on a particular case in yeast. There was even earlier work with *Aspergillus* (4). Dienert was a Pasteurian. Was he still working at the Institute? Could he be asked? Now, it is impossible to find out. However, I was told that in the late fifties a *very* old man whom nobody recognized came to Monod's office: he introduced himself as Dienert. He wanted to see Monod, the man who was solving in *Escherichia coli* the problem he had attacked in yeast: yeast grown on glucose lacks the enzymes which break down galactose, but yeast grown on galactose contains these enzymes.

Lwoff must have shown Monod the review on adaptation written in German by the Finnish biochemist Henning Karström (1899-1969) (5). There Monod saw a table (Table 1), which contained the major results of this part of his own thesis and which he duly reproduced. Bacteria grown on glucose cannot continue to grow on certain sugars, *e.g.* lactose, whereas bacteria grown on those sugars can continue to grow on glucose. Please note that Karström used the same carbon sources as Monod would use later. He also described similar, but less extensive experiments with *Escherichia coli*. He called the glucose degrading enzymes "constitutive" and the lactose, galactose, arabinose and maltose degrading enzymes "adaptive".

To study adaptation, Monod mainly used *Bacillus subtilis*, and to a lesser extent *E. coli*. Like Karström, he used all the carbon sources which he could find. First, he checked and confirmed that growth of both bacteria is strictly propor-

Table 1:   Constitutive and adaptive enzymes of a lactic acid bacterium (*Betacoccus*).

Bacteria	ferment					
grown in the presence of	Glucose Fructose Mannose	Galactose	Arabinose	Saccharose	Maltose	Lactose
Glucose	+	0	0	(+)	0	0
Galactose	+	+	0	+	0	0
Arabinose	+	0	+	+	0	0
Saccharose	+	0	0	+	0	0
Maltose	+	0	0	+	+	0
Lactose	+	+	0	+	0	+
no carbohydrate	+	0	0	+	(+)	0

+ = fermentation; 0 = no fermentation. Translated from Karström (5).

tional to the amounts of the carbon sources in the mineral medium. Then he confirmed the findings of Dienert (3) and Karström (5) that the microorganisms grow first in glucose, and then and only then on the other added carbon source (*i.e.* galactose, maltose etc.). This was demonstrated easily by using various amounts of glucose and of the other carbon source (Fig. 1). The more glucose is present, the higher the first plateau of growth. He also invented a catchy new name for this phenomenon – diauxie.

Before Monod began to work on this problem, a tentative explanation was given by John Yudkin (Cambridge, England). He postulated an equilibrium between a hypothetical pre-enzyme and the active enzyme and that binding of the active enzyme by its substrate displaced the equilibrium in the direction of more active enzyme. So, Yudkin proposed that enzymes folded around their substrates and that the speed of enzyme synthesis depended on substrate assisted folding, *i.e.* the presence of the substrates (6). This theory – Yudkin called it the *mass action theory* - was generally accepted. Karl Landsteiner adapted it without quoting it, to explain antibody production (7). The great Linus Pauling used it too, without quoting it as part of his detailed explanation for the synthesis of specific antibodies (8). The concept was also called the *instructive theory*.

I pause here to recall that Paris was occupied by the German Army from 1940 to 1944. It was a time of war. Monod finished his work on his thesis at the Sorbonne in 1941 (9). Then, he followed the example of one of his professors, Mar-

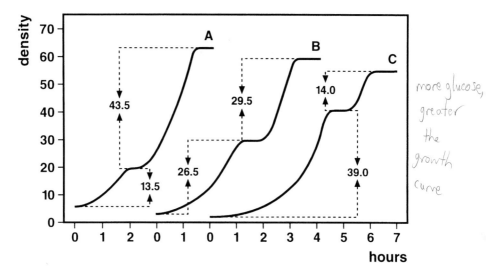

handwritten note in right margin: more glucose, greater the growth curve

Fig. 1.:   An example of diauxie (Fig. 51) taken from Monod's thesis (9). Growth of *E.coli* in a mixture of glucose and sorbite in the proportion 1/3 (A); 2/2 (B) and 3/1 (C). Abscissa: time in hours. Ordinate: bacterial density in relative units. The numbers inside the figure indicate total growth corresponding to each "growth cycle".

cel Prenant. He joined the illegal Communist party and went underground. Later, when asked about this period he remained rather silent (10).

Those were bitter years. I mention here only the fate of some of the people who later will appear in this book. I begin with Élisabeth and Eugène Wollman, a French couple working at the Institut Pasteur. They did research on lysogenic *Bacillus megatherium* and its phage (11). The phage and the bacteria behave exactly like λ and *E. coli*. Lysogenic bacteria are "immune" to attack by the phage. When grown in liquid media their culture supernatant always contains some phage which will attack non-lysogenic bacteria. This *was* the forerunner of the phage λ. The Wollmans were Jewish. So the Germans deported them to Auschwitz in 1944, where they were murdered. They had a son, Elie, who was 22 years old when the war began. He went underground and survived. His name will turn up later when I discuss bacterial mating. There was another young French Jewish student, Georges Cohen, who, like Monod, went underground and joined the Communist party. His name will turn up with Lac permease and *met* and *trp* control. And finally there was a very young medical student, François Jacob, also Jewish. He was just about to begin his medical studies when the war began. He went to England, joined the French army under De Gaulle, marched from Chad via Libya to Tunisia, participated on D-day in the landings in Normandy and was severely wounded there. When he saw his French classmates, who had studied during the war and who had taken their exams, he must have felt like a loser. He joined Lwoff's laboratory in 1950 (12).

References

(1) Lwoff, A.: The Prophage and I. In: Phage and the origins of molecular biology. Ed. by J. Cairns, G.S. Stent & J.D: Watson. Cold Spring Harbor Laboratory of Quantitative Biology, 88-99, 1966.

(2) Monod, J.: From enzymatic adaptation to allosteric transitions (Nobel lecture). Science **154**, 475-483, 1966.

(3) Dienert, F.: Sur la fermentation du galactose et sur l'accoutumance des levures à ce sucre. Ann.Inst.Pasteur **14**, 139-189, 1900.

(4) Duclaux, E.: Traité de microbiologie. Masson et Cie, Paris, 1899.

(5) Karström, H.: Enzymatische Adaption bei Mikroorganismen. Ergebnisse der Enzymforschung **7**, 350-376, 1938.

(6) Yudkin, J.: Enzyme variation in micro-organisms. Biol.Rev. **13**, 93-106, 1938.

(7) Rothen, A. & Landsteiner, K.: Adsorption of antibodies by egg albumin films. Science **90**, 65-66, 1939.

(8) Pauling, L.: A theory of the structure and process of formation of antibodies. J.Am.Chem.Soc. **62**, 2643-2657, 1940.

(9) Monod, J.: Recherches sur la croissance des cultures bactériennes (Thèse Doctorat ès Sciences) Hermann Ed., Paris, 1942.

(10) Judson, H.F.: The eighth day of creation. Makers of the revolution in biology. Simon and Schuster, New York, 1979.

(11) I quote here just two papers: Wollman, E. & Wollman, E.: Bactériophagie ou autolyse hérédo-contagieuse. Ann.Inst. Pasteur **56**, 137-170, 1936. Wollman, E. & Lacassagne, A.: Évaluation des dimensions des bactériophages au moyen des rayons X. Ann.Inst.Pasteur **64**, 5-39, 1940.

(12) Jacob, F.: La statue intérieure. Éditions Odile Jacob, Paris, 1987.

1.3 Mutants in the *lac* System of *Escherichia coli*

The tacit assumption of Monod's work had been that during diauxie all or most of the bacterial cells adapt quickly, within hours, to the new carbon source. Yet Karström also mentions (1) "slow adaptation by mutation in the sense of de Vries". He cites a 1907 paper by Rudolf Massini (2) who worked as a postdoc in the laboratory of Max Neisser at the university of Frankfurt, and who had isolated a variant of *E. coli* from human feces which was unable to grow on lactose. When Massini plated these bacteria on lactose indicator plates the colonies were MaConkey white (*lac⁻*). After some days, red papillae formed on the white colonies (when lactose is degraded, the pH changes in and around the bacteria and this changes the colour of a dye added to the plate and so present in and around the bacteria). Bacteria picked from the red papillae were *lac⁺*: they grew like ordinary *E. coli* on lactose, the rest of the colony remained *lac⁻*. He repeated this process many times, always with the same results. Thus, he came to the conclusion that the *lac⁺* bacteria were rare mutants of stable *lac⁻* bacteria. To differentiate these bacteria from ordinary *E. coli*, he called them *Bacterium coli mutabile*.

One would think that the subject of the analysis of bacterial mutants could now be attacked in a rational manner. This was far from so. Twenty four years later, I.M. Lewis, from the University of Texas, came to the conclusion (3): "The subject of bacterial variation and heredity has reached an almost hopeless state of confusion. Almost every possible view has been set forth and there seems no reason to hope that any uniform consensus of opinion may be reached in the immediate future. There are many advocates of a Lamarckian mode of bacterial inheritance while others hold the view that it is essentially Darwinian". Lewis himself again analysed *E. coli* mutabile strains, which could mutate from *lac⁻* to *lac⁺*. He showed convincingly that mutations from *lac⁻* to *lac⁺* occur spontaneously in the absence of lactose (3).

Some years later, Lwoff isolated a similar *E. coli* strain (ML3 = Mutabile Lwoffi, or according to knowledgable sources, more prosaically *merdae Lwoffi*). Monod and his student, Alice Audureau, repeated the old experiments of Massini and Lewis with these bacteria in 1946, and reproduced their results (4). When they plated about 5 x 10⁸ such bacteria on lactose plates, they found about 300 colonies which grew on lactose. If one grew such *lac⁺* mutants on a mixture of glucose and lactose, one observed the phenomenon of diauxie, as in ordinary *E. coli*. The bacteria grew first on glucose, and then and only then on lactose (4). One *lac⁺* revertant was kept and named ML30, and used for all subsequent ex-

[handwritten: ML3 Lac⁻ has no permease]

periments. The nature of the defect in the *lac* system was unknown. It took ten more years before it became clear that the *lac⁻* bacteria carry a defect in the lactose permease, the pump which transports lactose into the cells and the presence of which is essential for growth on lactose. Even today, the mutated *lac* DNA of ML3 has not been sequenced.

During the same year, 1946, Monod attended the conference at Cold Spring Harbor on Heredity and Variation in Microorganisms. There he met Max Delbrück and Salvadore Luria, who had analysed the nature of the *E. coli* mutation to *T1* resistance. They had shown that it involved a change in the genetic material which occurred by chance (5). Also present was Sol Spiegelman, who believed he had demonstrated that in yeast the precursors of the galactose metabolising enzymes multiplied during adaptation (6). Present, too, was Joshua Lederberg, who had shown in his doctoral thesis that one could isolate recombinants from crosses of *E. coli* strains carrying various auxotrophic mutations (7).

[handwritten: lacZ⁻]

One year later, Lederberg isolated mutants of *E. coli* which could not grow on lactose. Some were defective in β-galactosidase, the enzyme which breaks down lactose. They were called z^- (z is arbitrary) mutations. Other mutants had different unexplained defects (8). And perhaps even more startling, he isolated a mutant (9) which did not need to adapt to lactose, since it always produced large amounts of β-galactosidase. This mutant was called constitutive. In this mutant,

[handwritten: lacI⁻]

the mechanism which governs adaptation was clearly damaged. All these *lac* mutations appeared to be tightly linked; they recombined poorly when they were crossed to each other.

Both *lac⁻* and *lac* constitutive mutations are rare, spontaneous events. They occur about once in a million cell divisions. Thus it became important to devise techniques to *select* for such mutants. Such a technique for isolating *lac* constitutive mutants was designed in 1953 at the Institute Pasteur (10). During diauxie *lac* constitutive mutants have a growth advantage. If one uses several growth cycles, one strongly enriches for such *lac* constitutive mutants.

"Adaptation", the phenomenon Monod had set out to resolve, apparently consisted of two radically different phenomena: 1. the rapid increase of activity or concentration of an enzyme after "induction" with its substrate, the "fast adaptation" according to Karström; 2. the "selection of mutants" under suitable selection pressure, the "slow adaptation" according to Karström. Monod was a man who disliked sloppy language in science. He was also most apt at introducing new words to define phenomena. Thus together with Melvin Cohn, Martin Pollock, Sol Spiegelman and Roger Stanier he published a letter in Nature on the "terminology of enzyme formation" in 1953. There they proposed the use of the specific terms "induction" or "mutation" instead of the general term "adaptation", which meant both (11).

But what was induction? One has to keep in mind that before 1951 the nature of genes was unclear. It was the experiment of Alfred Hershey and Martha Chase which in 1951 changed people's minds. Phage *T2* was an object which one understood and where one accepted that DNA and not protein was the genetic material. Before 1951, Avery's paper (12) was cited but once by Monod (13), not as proof that the genetic material is DNA but as a rather special way of getting mutants! Protein, *i.e.* enzymes, was thought to be the genetic material. Enzymes possibly had the property of self-replication. Spiegelman thus proposed (6) that induction (to use Monod's terminology) was kicking on the self-replication of enzymes. The German biochemist Schoenheimer, who had gone into exile to the USA, spoke of the "dynamic state of living matter". Proteins seemed not to be stable, but in a process of constant turnover, new assembly and destruction. In fact a whole kinetic-mathematical theory was built on this Lamarckian concept by Sir Cyril Hinshelwood (14). We have to keep in mind that sequence identity of proteins was not known at this time. The immunologists believed that possibly all antibodies had the same sequence but just folded differently upon stimulation by antigen. Thus one could visualise inducers as instructing a protein to fold in a specific manner to become catalytically active.

Adaptation was the last stronghold of the Lamarckian view in which the surroundings influence living matter in a profound and long-lasting manner. In Germany, geneticists and the Nazis had presented a rather special antisemitic, Darwinian, anti-Lamarckian world view. According to them, human genetics determined the fate of humanity. In particular, the Nordic race, *i.e.* the race to which the Germans belonged, was superior. In contrast, the communists claimed that education (*i.e.* environment) is everything and that even plants and animals could be "educated". Their Soviet spokesman, Trofim Lysenko, claimed that Mendelian genetics was a fraud pushed by capitalists. Monod's teacher in biology at the Sorbonne and his direct superior in the underground Communist party, Marcel Prenant, remained a loyal communist when, in 1948, Mendelian genetics was formally damned in the Soviet Union (15). When the unrepenting Mendelian geneticists were thrown out of the Soviet Academy of Science, he defended this action in an article in the daily *Combat* (16). This was too much for Monod. One day later, he published his answer in the same daily under the headline: "the victory of Lysenko has no scientific character" (17). Monod then, like J.B.S. Haldane, H.J. Muller (18) and most other Western geneticists, left the Communist party or ended all connections. The Soviet Union was out, Lysenko was a fraud and that was that. Only Carl Lindegren, the teacher of Sol Spiegelman, dared to think aloud about Lamarckian mutations (19). About forty years later John Cairns has attempted a comeback for Lamarckian mutations (20). I will return to this later (section 2.21.).

References

(1) Karström, H.: Enzymatische Adaption bei Mikroorganismen. Ergebnisse der Enzymforschung **7**, 350-376, 1938.

(2) Massini, R.: Über einen in biologischer Beziehung interessanten Kolistamm (Bacterium coli mutabile). Arch.f.Hygiene **61**, 256-299, 1907.

(3) Lewis, I.M.: Bacterial variation with special reference to behavior of some mutabile strains of colon bacteria in synthetic media. J.Bact. **28**, 619-640, 1934.

(4) Monod, J. & Audureau, A.: Mutation et adaption enzymatic chez *Escherichia coli-mutabile*. Ann.Inst.Pasteur **72**, 868-879, 1946.

(5) Luria, S.E. & Delbrück, M.: Mutations of bacteria from virus sensitivity to virus resistance. Genetics **28**, 491-511, 1943.

(6) Spiegelman, S.: Nuclear and cytoplasmic factors controlling enzymatic constitution. In: Cold Spring Harbor Symposia on Quant.Biol.: Heredity and Variation in Microorganisms **11**, 256-277, 1946.

(7) Lederberg, J. & Tatum, E.L.: Novel genotypes in mixed cultures of biochemical mutants of bacteria. Cold Spring Harbor Symposia on Quant.Biol. **11**, 113-114, 1946.

(8) Lederberg, J.: Gene regulation and linked segregations in *Escherichia coli*. Genetics **32**, 505-525, 1947.

(9) Lederberg, J.: Genetic studies in bacteria. In: Genetics in the 20th Century. Macmillan, New York, 263-290, 1951.

(10) Cohen-Bazire, G. & Jolit, M.: Isolement par sélection de mutants d'*Escherichia coli* synthétisants spontanément l'amylomaltose et la β-galactosidase. Ann.Inst.Pasteur **84**, 937-945, 1953.

(11) Cohn, M., Monod, J., Pollock, M.R., Spiegelman, S. & Stanier, R.Y.: Terminology of enzyme formation. Nature **172**, 1096-1097, 1953.

(12) Avery, O.T., McLeod, C.M. & McCarthy, M.: Studies on the chemical nature of the substance inducing transformation of pneumococcal types. J.Exp.Med. **79**, 137-158, 1944.

(13) Monod, J.: The phenomenon of enzymatic adaptation and its bearing on problems of genetics and cellular differentiation. In: Growth Symposium XI, 223-289, 1947.

(14) Hinshelwood, C.N.: The chemical kinetics of the bacterial cell. Oxford, Clarendon Press, 1946.

(15) The situation in biological science. Proceedings of the Lenin Academy of Agricultural Sciences of the USSR. Sessions: July 31 – August 7, 1948. Verbatim Report. Foreign Languages Publishing House, Moscow, 1948.

(16) Prenant, M.: Interview. Combat, September 14, 1948.

(17) Monod, J.: "La victoire de Lyssenko n'a aucun caractère scientifique" estime le docteur Jacques Monod. Combat, September 15, 1948.

(18) It is totally beyond the scope of this book, but I urge the reader to read the books and articles by Muller and Haldane, for example: J.B.S. Haldane: Daedalus or Science and the future. Kegan Paul, Trench, Trubner & Co, London, 1924. Muller, H.J.: Out of the night. A biologists view of the future. Vanguard Press, New York, 1935. Muller, H.J.: Studies in genetics. The selected papers of H.J. Muller. Indiana University Press, Bloomington, Ind., 1962.

(19) Lindegren, C.C.: Cold war in biology. Planarian. Ann Arbor, Michigan, 1966.

(20) Cairns, J., Overbaugh, J. & Miller, S.: The origin of mutants. Nature **335**, 142-145, 1991.

1.4 Inducers of the *lac* System and the Discovery of Permease

By 1950, the *lac* system was one of half a dozen inducible systems (*ara*, *mal*, *trp*, *met*, *arg*) in *E. coli* which were equally well analysed. The *B. subtilis* systems could not compete, since there was no evidence for the sexual crosses which could be done in *E. coli*. The *lac* system had the same drawback as had all other systems: there was no fast and easy test of β-galactosidase (or lactase as it was then called), the enzyme which splits lactose into glucose and galactose. Furthermore, there was only one known inducer, lactose, which was broken down as the substrate of β-galactosidase. However, it was desirable to compare the specificity of β-galactosidase to the specificity of induction with a set of chemical analogues of the substrate and the inducer.

Until 1950 the action of β-galactosidase could be measured in two ways *in vitro*: the change in optical rotation upon breakdown of lactose into glucose and galactose could be determined. This was a laborious test which needed gram amounts of lactose and large amounts of enzyme. Alternatively, the oxidation of the breakdown product glucose could be measured in smaller samples, with Warburg's method. Again, this was laborious. It came as a relief when two American chemists followed the suggestion of Joshua Lederberg and synthesized o-nitrophenyl-β-D-galactoside (ONPG) as a substrate for β-galactosidase (1). ONPG is cheap and colourless. Upon attack by β-galactosidase it breaks down to colourless galactose and yellow o-nitrophenol. The yellow colour can be determined quantitatively with a cheap colorimeter. The test is so sensitive that one molecule of β-galactosidase per cell can be accurately determined in a few milliliters of cell suspension. Furthermore, the existence of *lac* constitutive *E. coli* strains made it easy to isolate gram amounts of β-galactosidase for detailed studies. Thus, β-galactosidase was analysed in great detail (2).

Monod collected all glycosides available in the Pasteur which were structurally similar to lactose and tested them as possible substrates and inducers of the *lac* system (3). The list of the compounds he analysed contains phenyl-1-thio-β-D-galactoside. This galactoside was not commercially available. It had been synthesized by Melvin Cohn in Cambridge (England), in analogy to a thio-glucoside which Monod had found in a drawer at the Pasteur. How did this thio-glucoside get in this drawer? Chemicals have their histories.

Phenyl-β-D-1-thio-glucoside had been synthesized in Berlin in the laboratory of Emil Fischer in 1909 by Konrad Delbrück, an uncle of Max Delbrück (4). Fischer (1852-1919) was one of the greatest, if not the greatest, organic chemist of his time. He was profoundly interested in biochemistry. He coined the metaphor of substrate and enzyme as lock and key in 1894 (5). For him the enzyme was the key which opened the lock of the substrate. How did the phenyl-β-D-1-thio-glucoside move from Berlin university to the Institut Pasteur in Paris? One of the many collaborators of Fischer, the later Nobel prize winner Otto Meyerhof (1884-1951), took a collection of Fischer's compounds with him when he became director of a *Kaiser Wilhelm* Institute in Heidelberg. He also took the collection with him when he was forced to leave Heidelberg in 1938 because he was a Jew. He fled to Paris where he obtained a position in the Pasteur. He left the chemicals there when he continued his flight via Marseille and Spain to the US in 1940 after the German victory over France.

Phenyl-1-thio-β-D-galactoside has most provocative properties (3). In contrast to phenyl-β-D-galactoside, it is not hydrolysed by β-galactosidase (in fact its hydrolysis is slowed down by a factor of about a million (6)). But it is a strong competitive inhibitor of lactose or ONPG hydrolysis by β-galactosidase. Yet it is *not* an inducer. This suggested that the specificity of induction and of β-galactosidase are different. Thus the real inducer does not necessarily interact with pre-β-galactosidase, but with another unknown molecule of the "inducing machinery".

The existence of phenyl-1-thio-β-D-galactoside inspired Monod to look for a chemist who would synthesize additional thio-galactosides, particularly alkyl derivatives. He found in Burckhardt Helferich of Bonn university a professor willing to put a graduate student, Dietmar Türk, on the problem. Türk synthesized a set of thio-galactosides (7) and Monod tested them immediately. The results were most rewarding. Two of the new thio-galactosides, the methyl- and the isopropyl-derivatives (the former was abbreviated TMG, the later IPTG) were excellent "gratuitous" (8) inducers, yet were not themselves hydrolysed by β-galactosidase.   ↖ inducers, but not substrate

When these compounds became commercially available, the kinetics of induction was analysed by others (9-11). If one used a high concentration of IPTG (5×10^{-3}M), induction began three minutes after adding the inducer (9-11). These three minutes was the time it took to synthesize the first molecules of β-galactosidase. And most important, the rate of synthesis of induced β-galactosidase remains constant during induction (8), see Fig. 2. This excludes the possibility that inducer interacts with a pre-enzyme to force it into the proper, active configuration. Were this the case, the rate of synthesis of β-galactosidase should decrease with time. These experiments were done with normal, unsyn-

↖ inducers don't interact w/ pre-enzyme they interact w/ enzyme

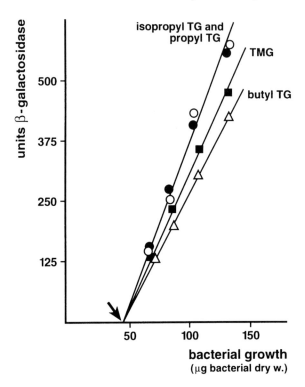

Fig. 2.:   Synthesis of ß-galactosidase in the presence of various alkyl thio-galactosides. Exponentially growing culture of *E.coli* ML 30 $i^+z^+y^+$ on mineral salts + succinate as sole
carbon source. Addition of inducers at point indicated by arrow (8).

chronised cultures of *E. coli*. Richard Goldberg, graduate student of Erwin
Chargaff, did the experiment with synchronised cells of *E. coli* (12). He observed
slight discontinuities connected with cell division. The only reason to quote this
rather uninteresting experiment is to alert the reader that Erwin Chargaff did not
only discover the rules about DNA composition (A=T; G=C). He is also the author of poems, aphorisms, a play, essays and an autobiography which are beautiful, intelligent and worth reading (13).

For those not convinced by the kinetics, the *de novo* synthesis of induced β-
galactosidase was demonstrated by two experiments. 1. Antibodies raised against
β-galactosidase were used as reagent (14). The existence of cross-reacting material called P_z which is always present in *E. coli* seemed first to argue for the
instructive theory. However, a careful quantitative analysis proved definitely *de*

novo synthesis (to the present day the gene which produces P_z has not been identified). 2. Radioactive $^{35}SO_4^-$ or ^{14}C labelled L lactic acid was added to the growth medium (15,16). Again it could be demonstrated that the induced β-galactosidase is not already partially present as protein before induction. Moreover it could be shown that β-galactosidase is stable and has no turnover after induction (17).

One could follow induction of enzymatically inactive mutant β-galactosidase with antibodies against β-galactosidase as reagent (18). There, one could, of course, still argue that the mutant inactive β-galactosidase still carries the binding site for the inducer. Catalytic inactivity does not imply the absence of a binding site. Among the thio-galactosides which Türk had synthesized was one, phenethyl-1-thio-β-D-galactoside, which is an excellent competitive inhibitor of β-galactosidase. It was used by Melvin Cohn to determine the number of binding sites, *i.e.* the active sites, in β-galactosidase. He reported one binding site per monomer (19). With this test the number of binding sites of the inactive β-galactosidase could be determined. No binding was observed. Thus, it was demonstrated that inactive mutant β-galactosidase is induced even when it lacks a proper binding site for inducer and substrate (18).

When the specificity of β-galactosidase and of the "inducing machinery" were compared, it was assumed that structure and concentration of the glycosides are the same both in the medium and inside the cell. To test this assumption, ^{35}S labelled, radioactive methyl-β-D-1-thio-galactoside (TMG) had to be synthesized. Dietmar Türk came from Bonn University to the Pasteur to do the job with 50 m curies of ^{35}S-methylmercaptan. He got help from Georges Cohen and Howard Rickenberg, an American postdoc (Rickenberg was born in Nuremberg as Hans Reichenberger. His parents sent him in 1937 alone to England to finish secondary school. There he was interned in 1940 and then deported to Australia. Later he emigrated to the USA). So, Cohen and Rickenberg tested the assumption in 1955. They showed that wildtype *lac⁺* cells accumulate radioactive methyl-β-D-1-thio-galactoside (TMG) about one hundredfold when it is offered at low concentration ($10^{-5}M$) in the medium (20). They showed further that this property is missing in the *E. coli* strain ML3 which Monod had mutated from *lac⁻* to *lac⁺* in 1946 (see section 1.3.). *E. coli* needs a specific pump with which to introduce lactose and other β-galactosides into the cell. Monod called this pump a permease (21). When the editor of Biochemica Biophysica Acta rejected the manuscript because of this name, Monod submitted it to the Annales de l'Institut Pasteur. There it was accepted.

Name	Structure	Properties

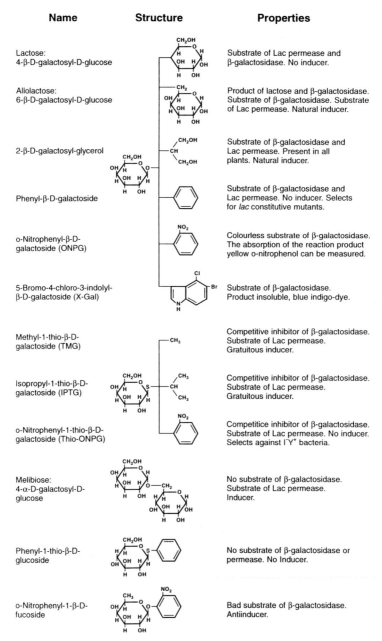

Lactose:
4-β-D-galactosyl-D-glucose

Substrate of Lac permease and β-galactosidase. No inducer.

Allolactose:
6-β-D-galactosyl-D-glucose

Product of lactose and β-galactosidase. Substrate of β-galactosidase. Substrate of Lac permease. Natural inducer.

2-β-D-galactosyl-glycerol

Substrate of β-galactosidase and Lac permease. Present in all plants. Natural inducer.

Phenyl-β-D-galactoside

Substrate of β-galactosidase and Lac permease. No inducer. Selects for *lac* constitutive mutants.

o-Nitrophenyl-β-D-galactoside (ONPG)

Colourless substrate of β-galactosidase. The absorption of the reaction product yellow o-nitrophenol can be measured.

5-Bromo-4-chloro-3-indolyl-β-D-galactoside (X-Gal)

Substrate of β-galactosidase. Product insoluble, blue indigo-dye.

Methyl-1-thio-β-D-galactoside (TMG)

Competitive inhibitor of β-galactosidase. Substrate of Lac permease. Gratuitous inducer.

Isopropyl-1-thio-β-D-galactoside (IPTG)

Competitive inhibitor of β-galactosidase. Substrate of Lac permease. Gratuitous inducer.

o-Nitrophenyl-1-thio-β-D-galactoside (Thio-ONPG)

Competitice inhibitor of β-galactosidase. Substrate of Lac permease. No inducer. Selects against I⁻Y⁺ bacteria.

Melibiose:
4-α-D-galactosyl-D-glucose

No substrate of β-galactosidase. Substrate of Lac permease. Inducer.

Phenyl-1-thio-β-D-glucoside

No substrate of β-galactosidase or permease. No Inducer.

o-Nitrophenyl-1-β-D-fucoside

Bad substrate of β-galactosidase. Antiinducer.

Fig. 3.:   Names, structures and properties of some glycosides mentioned in the text.

Most of this happened before 1957, at a time when the nature of the "inducing machinery" was totally unclear. I include here three additional discoveries which were made about ten years later. They amplify the picture without changing it.

- First, there is a sugar, o-nitrophenyl-β-D-fucoside (fucose is 6-deoxygalactose), which inhibits induction by IPTG at the level of the "inducing machinery" (22). Thus, we have inducers and anti-inducers, *i.e.* inhibitors of induction.
- Second, using an inducible (I^+) (I stands for inducible), galactosidase negative (Z^-) and Lac permease positive (Y^+) strain (Z stands for β-galactosidase, Y stands for permease), it was shown that lactose itself is not an inducer (22,23). Lactose is metabolized in part by β-galactosidase to its isomer 1-6-β-D-galactosido-glucose (allolactose). Indeed, allolactose is the natural inducer!
- Third, a β-galactoside was synthesized which is itself colourless but when hydrolysed produces an insoluble blue product (24). This 5-bromo-4-chloro-3-indolyl-β-D-galactoside (X-gal) colours blue bacterial colonies or animal tissues which contain β-galactosidase.
- Finally I cite as a curiosity the synthesis of 6-acetyl-ONPG which is neither a substrate of β-galactosidase nor an inducer (25). Why? It was synthesized by a student of Jeshajahu Leibowitz (1903-1994). Leibowitz is a most interesting writer, who understood both science and the Jewish religion. He sat often between all chairs and his books (26) should be read.

References

(1) Seidman, M. & Link, K.P.: Substituted salicylaldehydes and derivates. J.Am.Chem.Soc. **72**, 4324-4325, 1950.

(2) Wallenfels, K. & Malhotra, O.P.: Galactosidases. Advances Carbohydrate Chem. **16**, 239-298, 1961.

(3) Monod, J., Cohen-Bazire, G. & Cohn, M.: Sur la biosynthese de la β-galactosidase chez *Escherichia coli*. La specificité de l'induction. Biochemica et Biophysica Acta **7**, 585-599, 1951.

(4) Fischer, E. & Delbrück, K.: Über Thiophenol-glukoside. Chemische Berichte **42**, 1476-1482, 1909.

(5) Fischer, E.: Einfluß der Configuration auf die Wirkung der Enzyme I. Chemische Berichte **27**, 2985-2993, 1894.

(6) Müller-Hill, B.: Über Struktur und Wirkungsmechanismus der Alkoholdehydrogenase aus Bäckerhefe und der β-Galactosidase aus E. coli ML309. Doctoral thesis, Universitat Freiburg i.Br., 1962.

(7) Türk, D.: Über die Darstellung von Thiogalactosiden und Thioglucuroniden. Doctoral thesis, Universität Bonn, 1955.

(8) Monod, J.: Remarks on the Mechanism of Enzyme Induction. In: Enzymes: Units of Biological Structure and Function, Henry Ford Hospital International Symposium, Detroit. New York, Academic Press, 7-28, 1956. Türk's and Helferich's compounds are extensively used in this review. They are neither here nor elsewhere coauthors. Their names appear in a footnote.

(9) Novick, A. & Weiner, M.: The kinetics of β-galactosidase induction. In: A Symposium on Molecular Biology. Ed. by R.E. Zirkle. Chicago, University of Chicago Press, 78-90, 1959.

(10) Boezi, J.A. & Cowie, D.B.: Kinetic studies of β-galactosidase induction. Biophys.J. **1**, 639-647, 1961.

(11) Kepes, A.: Kinetics of induced enzyme synthesis. Determination of the mean life of galactosidase-specific mRNA. Biochem. Biophys.Acta **76**, 293-309, 1963.

(12) Goldberg, R.B. & Chargaff, E.: On the control of the induction of β-galactosidase in synchronous cultures of *Escherichia coli*. Proc.Natl. Acad.Sci. USA **68**, 1702-1706, 1971.

(13) Chargaff, E.: Gedichte, Privatdruck; Klett-Cotta, Stuttgart 1985; Bemerkungen, Klett-Cotta, 1981; Vorläufiges Ende, Klett-Cotta, 1990; Abscheu vor der Weltgeschichte, Klett-Cotta, 1988; Zeugenschaft, Klett-Cotta, 1985; Vermächtnis, Klett-Cotta, 1992; Ein zweites Leben, Klett Cotta, 1995; Heraclitean Fire, The Rockefeller University Press, New York, 1978.

(14) Cohn, M. & Torriani, A.M.: The relationship in biosynthesis of the β-galactosidase- and PZ-proteins in *Escherichia coli*. Biochim.Biophys.Acta **10**, 280-289, 1953.

(15) Hogness, D.S., Cohn, M. & Monod, J.: Studies on the induced synthesis of β-galactosidase in *Escherichia coli*: The kinetics and mechanism of sulfur incorporation. Biochim.Biophys.Acta **16**, 99-116, 1955.

(16) Rotman, B. & Spiegelmann, S.: On the origin of the carbon in the induced synthesis of β-galactosidase in *Escherichia coli*. J.Bact. **68**, 419-429, 1954.

(17) Rickenberg, H.V., Yanofsky, C. & Bonner, D.M.: Enzymatic deadaptation. J.Bact. **66**, 683-687, 1953.

(18) Perrin, D., Jacob, F. & Monod, J.: Biosynthese induite d'une protéine génétiquement modifiée, ne presentant pas d'affinité pour l'inducteur. Comptes Rendus Acad.Sci. **251**, 155-157, 1960.

(19) Cohn, M.: Contributions of studies on the β-galactosidase of *E. coli* to our understanding of enzyme synthesis. Bact.Rev. **21**, 140-168, 1957.

(20) Cohen, G.N. & Rickenberg, H.V.: Étude directe de la fixation d'un inducteur de la β-galactosidase par les céllules d'*Escherichia coli*. Comptes Rendus Acad.Sci. **240**, 466-468, 1955.

(21) Rickenberg, H.V., Cohen, G.N., Buttin, G. & Monod, J.: La galactoside-perméase d'*Escherichia coli*. Ann.Inst.Pasteur **91**, 829-857, 1956.

(22) Müller-Hill, B., Rickenberg, H.V. & Wallenfels, K.: Specificity of the induction of the enzymes of the Lac operon in *Escherichia coli*. J.Mol.Biol. **10**, 303-318, 1964.

(23) Burstein, C., Cohn, M., Kepes, A. & Monod, J.: Rôle du lactose et de ses produits métaboliques dans l'induction de l'opéron lactose chez *Escherichia coli*. Biochem.Biophys.Acta **95**, 634-639, 1965.

(24) Horwitz, J.P., Chua, R., Curby, R.J., Tomson, A.J., DaRooge, M.A., Fisher, B.E., Mauricio, J. & Klundt, I.: Substrates for cytochemical demonstration of enzyme activitiy. I. Some substituted 3-indolyl-β-D-glycopyranosides. J.Med.Chem. **7**, 574-575, 1964.

(25) Frohwein, J.Z.: Enzyme action on fully and partially acetylated β-D-galactosides. Enzymologia Acta Biocatalytica **25**, 297-307, 1963.

(26) Leibowitz, J.: Gespräche über Gott und die Welt mit Michael Shashar. (Hebr. Keter Publishing, Jerusalem) Dvorah Verlag, Frankfurt a.M. 1990. Leibowitz, J.: Vorträge über die Sprüche der Väter. (Hebr. Schocken Editions Tel Aviv) Context Verlag, Obertshausen 1984.

1.5 Negative Control through Repressor

In 1957 it was evident that the "inducing machinery" in the *lac* system differed from the enzyme which was induced, β-galactosidase. But how did the "inducing machinery" work? What was its molecular structure? There existed mutants in which the "inducing machinery" is altered or destroyed, the constitutive mutants. They produce, even in the absence of inducer, amounts of β-galactosidase and Lac permease which normally are produced only in the presence of high concentrations of the most effective inducer. The amounts of β-galactosidase could be quantified. In the absence of inducer statistically about three molecules of tetrameric β-galactosidase are present in each *E. coli* cell. In contrast *lac* constitutive I^- cells produce about 3,000 molecules of β-galactosidase per cell (1).

One can rephrase the question: what is destroyed in the *lac* constitutive cells? In September 1957 Monod gave a semi-private lecture on the *lac* system to a guest of the Institut Pasteur, Leo Szilard. He ended his lecture with the question: how is the *lac* system controlled? "By negative control of course", was Szilard's immediate answer. Continuing on he explained that there are two modes of control, negative and positive control, and that negative control is easier to achieve than positive control. It is easier to inhibit the start of a reaction than to provoke its start. Negative control could be counteracted by inducer. Thus in the presence of inducer, negative control could be abolished.

I digress here for a moment to say a few words about Leo Szilard (1898-1964). Szilard was a Hungarian-Jewish physicist. As a young man he worked in Fritz Haber's *Kaiser Wilhelm* Institute in Berlin-Dahlem and was far-sighted enough to leave Berlin *before* Hitler came to power (2). He went to England and organized there help for Jewish scientists who had to flee Germany. He had the idea of the nuclear chain reaction before Otto Hahn showed it to exist. He helped to write the letter Albert Einstein sent to US President Roosevelt in favour of constructing the atom bomb. He tried to stop the bomb being dropped on civilians, and left physics after the bombing of Hiroshima and Nagasaki to begin a campaign against the proliferation of nuclear weapons. He published several science fiction stories (3). In one story, an extra-terrestrial expedition of scientists excavates Central Station in New York City after an atomic war has destroyed all human life on earth. The pay toilets with their coin deposit boxes start a controversy. The older, established scientists believe that the coins were deposited as part of a religious cult. A young, iconoclast scientist proposes the hypothesis that the economic system of Terra demanded payments for everything, even for toilet use, and that this extreme form of capitalism ruined the civilisation of Terra. The young man is ridiculed by his elders for proposing a clever but absurd idea. It should also be mentioned that Szilard published his version of the Ten Commandments for scientists (2).

Monod immediately proposed an experiment to test Szilard's idea of "negative control". If it is true, then presence of negative control, the inducible wild-type (I^+), should be dominant over absence of control (I^-). This could be tested by performing a bacterial cross. Luigi Cavalli (4) and later William Hayes (5) had shown that the crosses Joshua Lederberg had described were limited to a very few *E. coli* cells in which an omnipresent *F*-factor (*F* for fertility) had become stably integrated into one of several possible regions of the circular *E. coli* chromosome. Lederberg's crosses did not produce a linear map of genes because his bacterial cultures contained mixtures of different integrated *F*-factors. One clone of such an integrated *F*-factor was called an *Hfr* strain (*High frequency of recombination*). Elie Wollman and François Jacob had studied the kinetics of the mating by interrupting it with a Waring blender. So they could demonstrate the linear transfer of the DNA from the donor to the acceptor cell (6,7). The crosses indicated that the genes for the inducing machinery (*I*), for β-galactosidase (*Z*) and for permease (*Y*) are closely linked but independent (8).

Monod thus proposed to use a particular *Hfr* strain for the cross. It transferred the *lac* system early and with high efficiency to produce transient diploids. If this *Hfr* strain of the genotype $I^- Z^-$ *smS* (streptomycin sensitive) was crossed with a female strain which was wildtype $I^+ Z^+$ *smR* (streptomycin resistant) in the presence of streptomycin, the low amount of β-galactosidase present did not increase. But if an *Hfr* $I^+ Z^+$ *smS* strain was crossed with an $F^- I^- Z^-$ *smR* strain in the presence of streptomycin, β-galactosidase was synthesized during the first half hour at a high rate, but from then on at a low rate in the recipient female cells. If one would observe just the first half hour of the transfer experiment, one would conclude that I^- is dominant over I^+. But if one waits long enough, *i.e.* half an hour longer, one observes the beginning of repression. This can be explained as follows: It takes some time for the transfered *I* gene to express its product, the "repressor", which later "represses" β-galactosidase synthesis. Repression is dominant. But it only begins to work after about half an hour (8).

[handwritten margin note: repression dominant but takes time]

Szilard is not listed as a co-author of the PArdee, JAcob, MOnod (PaJaMo) (8) paper. He is reported to have said that he found his idea too trivial to be honoured by coauthorship. The acknowledgement notes: "We are much indebted to Professor Leo Szilard for illuminating discussions during this work …". A year later, in 1960, Szilard published his own paper on "the control of formation of specific proteins in bacteria and in animal cells" (9). He stuck to negative control. He had heard from Monod that there was evidence that Lac repressor was RNA. Being neither a chemist nor an immunologist, he represented the repressor-inducer complex as a RNA-inducer hybrid and proposed that antigens induced the synthesis of antibodies by binding to their mRNA.

How general is the phenomenon of repression? The weight of the PaJaMo (often pronounced *Pyjama*) paper (8) was increased considerably by the previous discovery of other, similar systems such as the *trp* (10) (the tryptophan), the *met* (11) (the methionine) and the *arg* (12) (the arginine) systems, the first two by Georges Cohen at the Pasteur. In fact, the term "repression" had been coined by Henry Vogel to describe the down-regulation of the synthesis of these systems by their metabolites (13). But the most spectacular similarity occurs in the phage λ system.

Here I again digress for a moment. Phage λ was discovered in 1953 by Esther Lederberg (14), then the wife of Joshua Lederberg. Phage λ develops in, and co-exists, as a lysogen in *E. coli* K12, the *E. coli* strain her former husband had used to demonstrate mating in bacteria. It was very much like the system the Wollmans had used in *B. megatherium*. In fact, Lwoff and his collaborators had shown in 1950 that lysogenic *B.megatherium* could be induced by exposure to UV light (15). This was a breakthrough, both in understanding and technique. It opened the possibility of obtaining large amounts of phage. But why continue with this phage in *B.megatherium*, a bacterium about which little was known, when a similar phage was around in *E. coli*? François Jacob wrote his thesis on lysogenic bacteria and the fact that *E. coli* lysogenic for phage λ could be induced by UV light like their counterparts in *B.megatherium* (16). To the best of my knowledge, since then the *B.megatherium* system has been abandoned. The first sentence in chapter one of Mark Ptashne's excellent book (17) about phage λ attests to the disappearance of the Wollmans and their phage from scientific memory. λ was victorious, who cares about its predecessor!

Jacob had the lucid idea of comparing phage λ with the *lac* system. He describes in his autobiography how the idea hit him that the "immune" system of λ and the regulatory system of *lac* were similar. It happened in August of 1958 when everybody was on vacation out of Paris (18). He was watching a Western movie together with his wife, when the idea overwhelmed him. The λ *CI* clear mutants could be compared to the *lac* constitutive, repressor-negative mutants. The absence of repressor led in one case to the production of phage proteins, in the other case to the production of β-galactosidase! He had nobody with whom to talk. It took several weeks until he could tell his story to Monod. The first day Monod ridiculed him, the second day he believed him. A few days later, Monod had a most interesting idea to test Jacob's proposition: as there are constitutive mutants in the repressors, there should also be constitutive mutants in the target of the repressors which presumably is DNA itself. These mutants would be *cis* dominant. Then, only then, it occurred to Jacob that five years earlier he had given a talk on such mutants in λ at a Cold Spring Harbor conference (19)! He had called them λ *v* (*v* stands for virulent) mutants. They multiply in the presence of λ *CI* repressor produced by a lysogen, *i.e.* they give plaques on lysogenic bacteria. Jacob and Monod gave the target of repressor a name. They called it operator, and published a short note about it (20). So Jacob looked for the analogous, *cis* dominant, operator constitutive mutants in the *lac* system.

[handwritten margin notes: "λ muts produce lots of proteins"; "repressors bind operators"]

References

(1) Wallenfels, K. & Malhotra, O.P.: Galactosidases. Advances in Carbohydrate Chem. **16**, 239-298, 1961.

(2) Szilard, L.: His version of the facts. Selected Recollections and Correspondence. Ed. by S.R. Weart and G. Weiss Szilard. MIT Press, Cambridge Mass. and London UK, 1978.

(3) Szilard, L.: Central Station. In: The voice of the dolphins. Simon and Schuster, New York, 1961.

(4) Cavalli-Sforza, L.L.: La sessualita nei batteri. Boll.Ist.sierotera Milanese **29**, 281-289, 1950.

(5) Hayes, W.: The mechanism of genetic recombination in *E. coli* K-12. Cold Spring Harbor Symposium on quant.Biol. **18**, 75-94, 1953.

(6) Wollman, E. & Jacob, F.: Sur le méchanisme du transfert de matériel génétique aux cours de la recombination chez *Escherichia coli*. Comptes Rendus Acad.Sci. **240**, 2449-2451, 1955.

(7) Jacob, F. & Wollman, E.L.: Sexuality and the Genetics of Bacteria. New York, Academic Press, 1961.

(8) Pardee, A., Jacob, F. & Monod, J.: The genetic control of cytoplasmic expression of "inducibility" in the synthesis of β-galactosidase by *E. coli*. J.Mol.Biol. **1**, 165-178, 1958.

(9) Szilard, L.: The control of the formation of specific proteins in bacteria and in animal cells. Proc.Natl.Acad.Sci. USA **46**, 277-292, 1960.

(10) Cohen-Bazire, G. & Monod, J.: L'effect inhibition spécifique dans la biosynthèse de la tryptophanedesmase chez *Aerobacter aerogenes*. Comptes Rendus Acad.Sci. **236**, 530-532, 1953.

(11) Cohn, M., Cohen, G.N. & Monod, J.: L'effect inhibiteur spécifique de la méthionine dans la formation de la méthionine-synthetase chez *Escherichia coli*. Comptes Rendus Acad.Sci **236**, 746-748, 1953.

(12) Gorini, L. & Maas, W.R.: The potential for the formation of a biosynthetic enzyme in *Escherichia coli*. Biochim.Biophys. Acta **25**, 208-209, 1957.

(13) Vogel, H.: Repression and induction as control mechanisms of enzyme biogenesis: The "adaptive" formation of acetylo-ornithinase. In: The chemical basis of heredity. Ed. by W.D. McElroy and B. Glass. Baltimore, John Hopkins Press, 276-289, 1957.

(14) Lederberg, E.M.:Lysogenicity in *E. coli* K-12. Genetics **36**, 560, 1951.

(15) Lwoff, A., Siminovitch, L. & Kjeldgaard, N.: Induction de la production de bacterio-phages chez une bactérie lysogène. Ann.Inst.Pasteur **79**, 815-859, 1950.

(16) Jacob, F.: Les bactéries lysogènes et la notion de provirus, Paris, Masson et Cie, 1954.

(17) "Some 40 years ago, André Lwoff and his colleagues at the Institut Pasteur in Paris described a dramatic property of a certain strain of the common intestinal bacteria *Escherichia coli*. If irradiated with a moderate dose of ultraviolet light these bacteria stop growing and some 90 minutes later they lyse (burst), spewing a crop of viruses called λ into the culture medium." In: Ptashne, M.: A Genetic Switch. Gene control and Phage λ / Phage λ and higher organisms. Cell and Blackwell Scientific Publications. 1. edition 1986, 2. edition 1992.

(18) Jacob, F.: La statue intérieure. Éditions Odile Jacob. Paris, 1987. The statue within: an autobiography. Basic Books, New York, 1988.

(19) Jacob, F. & Wollman, E.L.: Induction of phage development in lysogenic bacteria. Cold Spring Harbor Symposia on Quant.Biol. **18**, 101-120, 1953.

(20) Jacob, F. & Monod, J.: Gènes de structure et gènes de régulation dans la biosynthese des proteines. Comptes Rendus Acad.Sci. **249**, 1282-1284, 1959.

1.6 Further Evidence for Negative Control through Repressor

The proof of negative control exerted by Lac repressor was well documented in the PaJaMo-paper of 1959 (1). The next year brought forth an explosion of additional, most important pieces of evidence. I will list them here and just note that they were published as short notes, two in French. And I add, what a time this was, when important contributions to molecular biology could be published in French and were read by a substantial part of the scientific community!

1. If repressor interacted with a particular region of the gene it regulated – the operator – then mutants of the operator region (O^c) should exist in which the interaction with Lac repressor is damaged. They should be constitutive and *cis* dominant. Such *cis* dominant O^c mutants were indeed found and described in a short paper with the title: "L'opéron: groupe de gènes à expression coordinée par un opérateur" (2). Please note the coining of the term *operon*!
2. Proof of the dominance of I^+ over I^- by *Hfr* matings and *unstable* heteromerozygotes (cells which carried in addition to the chromosome other single genes) was sufficient but not ideal. It came in handy that Jacob and Adelberg (3) had just isolated a *stable F'lac* episome which occurs in one to two copies per *E. coli* cell. Now the dominance could be tested with great precision in *stable* heteromerozygotes.
3. Lac permease could be quantified but it was not an enzyme. It was therefore convenient that Leonard Herzenberg, working in Georges Cohen's lab, showed that when Lac permease was tested *in vivo* with radioactive isopropyl-1-thio-β-D-galactoside (IPTG), some of the IPTG was acetylated. This happened only in cells which were *lac* constitutive or which had been induced (4). It was shown that the gene coding for this enzyme, transacetylase (A) was part of the lactose (*lac*) operon and mapped in the order *I O Z Y A*. (The names for *I*, *O*, *Z*, and *Y* have been explained. Why the gene of transacetylase was called *A* and not X is Monod's secret.)
4. If mutants exist in which λ *CI* repressor has lost the capacity to bind its operator (*i.e. CI* clear mutants), then mutants (*CIind*) should exist which have lost the capacity to be induced by UV irradiation. Such mutants were found by Jacob and Alan Campbell (5). Similarly, in the *lac* system mutants (I^s) should exist which have lost the capacity to bind inducer. They should be *lac* negative (they cannot be induced) and dominant over I^+ and I^- mutants. Such mu-

lacY
lacA

- Mutants that can't bind operator
" " inducer (I^s)

tants were found in 1960 by Jacob. Formally they were similar to the λ *ind* mutants which could not be induced by UV irradiation. One mutant is briefly mentioned in the two Jacob & Monod papers of 1961 (6,7) and presented in detail in 1964 (8).

The years of 1959 and 1960 were perhaps the most productive years in the scientific life of Monod. These were the years when the Jacob-Monod theory of gene regulation came into being. But Monod was not only a scientist but also a courageous, decent man. So, in 1960 he found the time to travel to Budapest and to smuggle out of Hungary two dissident students: Agnes Ullmann and her husband Tom. The reader will hear more about Agnes Ullmann's achievements when I discuss complementation, the promoter and the CAP protein.

References

(1) Pardee, A., Jacob, F. & Monod, J.: The genetic control and cytoplasmic expression of "inducibility" in the synthesis of β-galactosidase by *E. coli*. J.Mol.Biol. **1**, 165-178, 1959.

(2) Jacob, F., Perrin, D., Sanchez, C. & Monod, J.: L'opéron: groupe de gènes à expression coordinée par un opérateur. Comptes Rendus Acad.Sci. **250**, 1727-1729, 1960.

(3) Jacob, F. & Adelberg, E.A.: Transfer de caractères génétiques par incorporation au facteur sexuel d'*Escherichia coli*. Comptes Rendus Acad.Sci. **249**, 189-191, 1959.

(4) Zabin, I., Kepes, A. & Monod, J.: On the enzymic acetylation of isopropyl-β-D-thiogalactoside and its association with galactoside-permease. Biochem.Biophys.Res.Com. **1**, 289-292, 1959.

(5) Jacob, F. & Campbell, A.: Sur le système de répression assurant l'immunité chez les bactéries lysogènes. Comptes Rendus Acad.Sci. **248**, 3219-3221, 1959.

(6) Jacob, F. & Monod, J.: Genetic regulatory mechanisms in the synthesis of proteins. J.Mol.Biol. **3**, 318-356, 1961.

(7) Jacob, F. & Monod, J.: On the regulation of gene activity. In: Cold Spring Harbor Symposia of Quantitative Biology **26**, 193-211, 1961.

(8) Willson, C., Perrin, D., Cohn, M., Jacob, F. & Monod, J.: Non-inducible mutants of the regulator gene in the "lactose" system of *Escherichia coli*. J.Mol.Biol. **8**, 582-592, 1964.

1.7 The Triumph of the Jacob-Monod Theory

At the end of 1960 an overwhelming amount of supporting data for a very specific model of gene control had been produced by Jacob and Monod and their
collaborators at the Institut Pasteur. Just one piece of data did not fit: one inducer, inositol-β-D-galactoside, induced just Lac permease, but not β-galactosidase
or transacetylase (1)! This piece of evidence resisted becoming part of the operon theory to the bitter end. Courageously Monod decided to neglect it. It was
not even mentioned. This was the right decision. Just five years later it turned
out that it was not the Lac permease which was induced by this inducer but the
homologous permease of the melibiose (*mel*) system (2)! Both permeases transport both β- and α-galactosides!

The year 1961 became the year when the model was presented (3-4) and discussed (5) in three major publications. The model (see Fig. 4) can be stated in the
following way: in bacteria transcription, *i.e.* mRNA production (6,7), is regulated
in a negative manner by repressors. In some systems, like the *lac* system, inducer
(a small molecule) is needed to counteract the action of the repressor. The inducing agent can also be UV light as in phage λ. In other systems like the anabolic
tryptophan (*trp*) or arginine (*arg*) systems, a small molecule is needed to act as a
corepressor to make the repressor work as a repressor.

The negative role of repressor predicts that repressor (R^+) may mutate to constitutive, repressor negative (R^-) bacteria. Presence of repressor (R^+) is dominant
over the absence of repressor (R^-). Here, in addition to Lac, the examples of Trp
repressor (8) and of λ *CI* repressor (9) could be quoted.

The existence of a repressor necessitates the existence of a region on the chromosome (or the mRNA), the operator, with which it interacts in order to function.
Thus mutants should exist in which the binding of repressor to its operator does
not work properly because the operator is damaged. These mutants should be
constitutive. In contrast to the R^- mutants which are recessive, the operator constitutive mutants (O^c) should be *cis* dominant. Evidence for this was produced. I
reproduce here part of the relevant table (Table 2) from the paper in the *Journal
of Molecular Biology* (3).

Finally, Occam's razor was used to decide at which level control occurs. The
simplest, and for bacteria the most economical way is to control at the DNA level.
This became the model. All the statements made so far were correct. The model
could be easily understood. The terms used to describe it were well chosen. It was
an immense success.

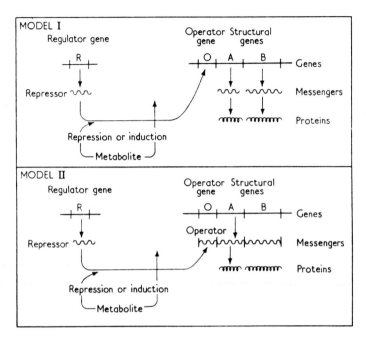

Fig. 4.:   The operon model of Jacob and Monod (3), with permission.

Table 2:   Synthesis of ß-galactosidase and transacetylase by haploid and heterozygous diploid
operator mutants.

	ß-Galactosidase		Transacetylase	
	Noninduced	Induced	Noninduced	Induced
$O^+Z^+Y^+$	<0.1	100	<1	100
$O^cZ^+Y^+$	25	95	15	110
$O^+Z^+Y^-/F'O^cZ^+Y^+$	70	220	50	160
$O^+Z^-Y^+/F'O^cZ^+Y^-$	180	440	<1	220

The Y^- mutant is a polar nonsense mutation. It reduces transacetylase activity drastically because of
the degradation of the untranslated part of the Y mRNA. All constructs are I^+. Adapted from (3).

References

(1) Jacob, F.: The Switch. In: Origins of molecular biology. A tribute to Jacques Monod. Ed. by A. Lwoff & A. Ullmann. Academic Press, New York, San Francisco, London. 95-107, 1979

(2) Prestidge, L.S. & Pardee, A.B.: A second permease for methyl-thio-β-galactoside in *E. coli*. Biochim.Biophys.Acta **100**, 591-593, 1965.

(3) Jacob, F. & Monod, J.: Genetic regulatory mechanisms in the synthesis of proteins. J.Mol.Biol. **3**, 318-356, 1961.

(4) Jacob, F. & Monod, J.: On the regulation of gene activity. In: Cold Spring Harbor Symposia of Quant.Biol. **26**, 193-211, 1961.

(5) Monod, J. & Jacob, F.: General conclusions: Teleonomic mechanisms in cellular metabolism, growth and differentiation. In: Cold Spring Harbor Symposia on Quant.Biol. **26**, 389-401, 1961.

(6) Brenner, S., Jacob, F. & Meselson, M.: An unstable intermediate carrying information from genes to ribosomes for protein synthesis. Nature **190**, 576-581, 1961.

(7) Gros, F., Gilbert, W., Hiatt, C., Kurland, C.G., Risebrough, R.W. & Watson, J.D.: Unstable ribonucleic acid revealed by pulse labelling of *Escherichia coli*. Nature **190**, 581-585, 1961.

(8) Cohen, G.N. & Jacob, F.: Sur la répression de la synthèse des enzymes intervenant dans la formation du tryptophane chez *Escherichia coli*. Comptes Rendus Acad.Sci. **248**, 3490-3492, 1959.

(9) Lwoff, A.: Lysogeny. Bact.Rev. **17**, 269-337, 1953.

1.8 Minor Defects in the Jacob-Monod Theory

The papers by Jacob & Monod in the *Journal of Molecular Biology* (1) and in the Cold Spring Harbor Symposium of 1961 (2) were giant steps forward. Suddenly, there existed a clear molecular picture of enzyme induction and repression. However it was inevitable that the model had some defects. They are listed here:

1. Both papers trusted the interpretation of a paper by Pardee & Prestidge (3) that the formation of a very few molecules of Lac repressor in the presence of an inhibitor of protein synthesis, chloramphenicol, proves that Lac repressor is not protein and thus demonstrates that it is RNA. Luria *et al.* (unpublished results) are quoted (2) with similar results in another system. Chemical common sense seems to have left the authors. Proteins were known to be able to bind specifically to almost any small organic molecule. Nothing comparable was known about RNA. How could they explain the specific, tight binding of various inducers to repressor RNA?

   In fact, only one year later Raquel Sussman, a collaborator of Jacob and Monod, provided evidence that λ *CI* repressor is a protein! Among 300 λ *CI* (clear) mutants 11 could be suppressed by a nonsense suppressor (4). It took three more years until similar experiments with similar results were done in the *lac* system. Constitutives isolated from cultures of I^S (5) or I^+ (6) bacteria were analysed in the absence and in the presence of nonsense suppression. Two suppressible nonsense mutants were found among 187 *lac* constitutives (6). Again the conclusion was: Lac repressor is a protein, not RNA.

2. Strong polar point mutations which map close to the operator and which knock out all activity of β-galactosidase, permease and transacetylase were misinterpreted as mutations of the operator, the DNA region which by definition interacts with repressor. These mutations were called O^o (operator zero) mutations. In 1961 the region which interacts with RNA-polymerase, the promoter (*P*), had as yet no name. This misinterpretation was alive for two years until it was disproved by the demonstration that some of the O^o mutants can be suppressed by amber nonsense suppressors (7). Up to fifty percent of the normal amounts of β-galactosidase, permease and may be produced in the presence of such a suppressor. Furthermore, the O^o mutations map outside the region defined by the O^c mutations but within the *lacZ* gene (8). The polarity of the nonsense mutations increases the closer the mutation is positioned to the 5' end of the *lacZ* gene. The non-translated part of the *lacZ* mRNA can be

attacked and degraded by various RNases downstream of the nonsense mutation. Thus the amount of permease and transacetylase may be drastically reduced by polar nonsense mutations.

Later, it was shown in the galactose (*gal*) system, which is regulated by a
repressor related to the Lac repressor, that some of its O^o mutations were due
to insertions of transposons (9,10). This was the beginning of the analysis of
bacterial transposons. Somewhat later the same transposons were found in the
lacZ gene (11).

3. The case for negative control was made so convincingly that positive control,
wherever it appeared, was assumed to be a mistake or error in analysis. Thus
positive control systems such as the arabinose (*ara*) (12,13) or the maltose
(*mal*) (14) system were greeted with suspicion and derision. Many, too many,
people, thought they had to be wrong. It took years until positive control obtained the recognition it deserved.

References

(1) Jacob, F. & Monod, J.: Genetic regulatory mechanisms in the synthesis of proteins.
    J.Mol.Biol. **3**, 318-356, 1961.
(2) Jacob, F. & Monod, J.: On the regulation of gene activity. Cold Spring Harbor Symposia
    on Quant.Biol. **26**, 193-211, 1961.
(3) Pardee, A.B. & Prestidge, L.S.: On the nature of the repressor of β-galactosidase synthesis
    in *Escherichia coli*. Biochem. Biophys.Acta **36**, 545-547, 1959.
(4) Jacob, F., Sussman, R. & Monod, J.: Sur la nature du répresseur assurant l'immunité des
    bacteries lysogènes. Comptes Rendus Acad.Sci. **354**, 4214-4216, 1962.
(5) Bourgeois, S., Cohn, M. & Orgel, L.: Suppression of and complementation among mu-
    tants of the regulatory gene of the lactose operon of *Escherichia coli*. J.Mol.Biol. **14**, 300-
    302, 1965.
(6) Müller-Hill, B.: Suppressible regulator constitutive mutants of the lactose system in *Es-
    chericha coli*. J.Mol.Biol. **15**, 374-375, 1966.
(7) Beckwith, J.R.: Restoration of operon activity by suppressors. Biochem.Biophys.Acta
    76, 162-164, 1963.
(8) Beckwith, J.R.: A deletion analysis of the *lac* operator region in *Escherichia coli*.
    J.Mol.Biol. **8**, 427-430, 1964.
(9) Shapiro, J.A.: Mutations caused by the insertion of genetic material into the galactose
    operon of *Escherichia coli*. J.Mol.Biol. **40**, 93-105, 1969.

(10) Jordan, E., Saedler, H., & Starlinger, P.: O° and strong-polar mutations in the gal operon are insertions. Molec.Gen.Genetics **102**, 353-363, 1968.

(11) Malamy, M.H.: Some properties of insertion mutations on the lac operon. In: The lactose operon. Ed. by J.R. Beckwith and D. Zipser. Cold Spring Harbor Laboratory, 359-373, 1970.

(12) Englesberg, E.: Discussion of the Jacob & Monod paper in the Cold Spring Harbor Symposia on Quant.Biol. **26**, 209-210, 1961.

(13) Englesberg, E., Irr, J., Power, J. & Lee, N.: Positive control of enzyme synthesis by Gene C in the *L-arabinose* system. J.Bact. **90**, 946-957, 1965.

(14) Schwartz, M.: Expression phénotypique et localisation génétique de mutations affectant le métabolisme du maltose chez *Escherichia coli* K12. Ann.Inst.Pasteur **112**, 673-702, 1967.

1.9 Isolation of the Lac and λ *CI* Repressors

, Knew were prot b/c there were nonsense supp

The existence of nonsense suppressor mutants in the λ *CI* and the *lacI* genes had shown that both repressors were proteins (1-3). The PaJaMo experiment suggested that Lac repressor is present at very low concentrations. If one made the reasonable assumption that ten molecules of Lac repressor, of a molecular weight of 30,000 Daltons, are present in each cell of *E. coli*, this corresponds to a repressor concentration of 10^{-8}M, or to about 0.002 percent of the total protein. The situation for λ *CI* repressor is similar. This seemed below the threshold of all conceivable tests.

The best possible *in vitro* test for Lac repressor seemed equilibrium dialysis with radioactive inducer. It was convenient that just at that time radioactive sulfur-labelled IPTG became commercially available. If one assumes that the concentration of IPTG which induces β-galactosidase synthesis twofold above the basal level (10^{-6}M) is identical to the dissociation constant of the repressor IPTG complex, one can estimate that the presence of Lac repressor cannot be demonstrated in a crude extract. An *E. coli* cell is presumably about 10^{-8} molar in Lac repressor, so Lac repressor has to be purified at least tenfold before it can be detected by equilibrium dialysis. A way out of this difficulty seemed to be the isolation of a mutant Lac repressor which binds IPTG tighter than does wildtype. Such a mutant should be induced at a concentration of IPTG lower than wildtype. But how could one isolate such a mutant?

Some years earlier, Novick & Weiner (4) and independently Cohn & Horibata (5), had shown that the *lac* system may act as a switch. If one grows *E. coli* cells, which are uninduced in the *lac* system, in the presence of 10^{-6}M IPTG, the cells may grow for a hundred generations, but will never become induced. If they have been induced before, they will remain induced on this low concentration of IPTG for more than a hundred generations. Once the Lac permease is turned on, it increases the intracellular IPTG concentration about one hundredfold. Thus, I tried to isolate *lacI* mutants which were induced at a lower concentration of IPTG. Such a mutant (I^t) (*t* for tight binder) was isolated and was made homozygous with an *F'lac* episome. Walter Gilbert showed that extracts from this mutant *E. coli* indeed indicate the presence of an IPTG binding protein, presumably Lac repressor. This protein was purified. It precipitated at 30 percent ammonium sulfate. It was absent in extracts of the I^- and I^s mutants. The dissociation constant of IPTG to I^t repressor was lower by a factor of about two compared to wildtype I^+ repressor. Lac repressor had been isolated (6). In retrospect we now recognize that an unexpected fact worked in our favour, viz. that the ten molecules of Lac repressor are tetrameric, *i.e.* contain four binding sites for IPTG.

At the same time, Mark Ptashne worked on the isolation of λ *CI* repressor in the same building of the Biolabs at Harvard. His scheme relied on the fact that there is a λ nonsense mutant (*Nsus*) which when introduced in *E. coli* produces only λ *CI* repressor and a λ protein called Rex, but no other λ proteins. Therefore he UV irradiated *E. coli* to destroy all functioning *E. coli* DNA, so that no more mRNAs could be made from *E. coli* DNA. Then he introduced λ *CI⁺Nsus* into one batch of *E. coli* and λ *CIsus Nsus* into another batch of *E. coli* (the *CIsus* contains a nonsense mutation in the gene coding for λ *CI* repressor). He added ^{3}H labelled amino acids to the first batch and ^{14}C labelled amino acids to the second batch. He let each batch produce *in vivo* active λ *CI* repressor, or its inactive, short nonsense fragment, respectively. He then lysed the bacteria and mixed the two extracts. If λ *CI* repressor were to be purified on a column, it should be present as a ^{3}H peak against the baseline of ^{3}H/^{14}C. He got a peak. It worked! Thus, λ *CI* repressor was isolated (7). (If the experiment had been done exactly as I describe it, it may not have worked. UV treatment of *E. coli* induces the production of a protein which degrades λ *CI* repressor, as we have seen in chapters 1.5 and 1.7. Fortunately, a mutant of λ exists, *ind*, which resists attack by the UV generated protein. So this λ *CI ind* mutant was used as starting material.)

Each technique used for repressor isolation was special. The technique used for Lac repressor could not be applied to any other system. The technique used for λ *CI* repressor was used for the isolation of the homologous phage *434* repressor (8). Geoffrey Zubay tried to introduce a general technique for repressor isolation. He set up an *in vitro* system for the synthesis of β-galactosidase starting from φ*80 lac* DNA (9). He used a gene fusion where a *trp* promoter-operator region was fused to a complete *lacZ* reporter gene as an indicator for the *trp* repressor-operator system (10). Similarly, this technique was also only used one more time for the isolation of Arg repressor (11). The invention of *in vitro* gene cloning made all these *in vivo* techniques too laborious, *i.e.* antique.

To isolate the first repressors was a challenge. Agnes Ullmann tried it in the Pasteur, Ethan Signer at the MIT. Others tried it elswhere. When it was all over, Monod came to visit Jim Watson at the Biolabs. The year before, together with François Jacob and André Lwoff he had received the Nobel Prize. At this time he gave a talk at Harvard Medical School on allostery, I had not attended; I was too busy. There he stood in the door of my lab and caught me in the midst of an experiment. He stared at me and said: "After all, Benno, it was pedestrian", and disappeared before I could answer. I have no witness for this story, Monod is dead. But it is known that he liked such jokes (12).

References

(1) Jacob, F., Sussman, R. & Monod, F.: Sur la nature du répresseur assurant l'immunité des bacteries lysogènes. Comptes Rendus Acad.Sci. **354**, 4214-4216, 1962.

(2) Bourgeois, S., Cohn, M. & Orgel, L.E.: Suppression of and complementation among mutants of the regulatory gene of the lactose operon of *Escherichia coli*. J.Mol.Biol. **14**, 300-302, 1965.

(3) Müller-Hill, B.: Suppressible regulator constitutive mutants of the lactose system in *Escherichia coli*. J.Mol.Biol. **15**, 374-375, 1966.

(4) Novick, A. & Weiner, M.: Enzyme induction, an all or none phenomenon. Proc.Natl.Acad.Sci. USA, **43**, 553-566, 1957.

(5) Cohn, M. & Horibata, K.: Physiology of the inhibition by glucose of the induced synthesis of the β-galactosidase-enzyme system of *Escherichia coli*. J.Bact. **78**, 624-635, 1959.

(6) Gilbert, W. & Müller-Hill, B.: Isolation of the *lac* repressor. Proc.Natl.Acad.Sci. USA **56**, 1891-1898, 1966.

(7) Ptashne, M.: Isolation of the λ phage repressor. Proc.Natl.Acad.Sci. USA **57**, 306-313, 1967.

(8) Pirotta, M. & Ptashne, M.: Isolation of the 434 phage repressor. Nature **222**, 541-544, 1969.

(9) Zubay, G. & Lederman, M.: DNA-directed peptide synthesis. VI Regulating the expression of the *lac* operon in a cell free system. Proc.Natl.Acad.Sci. USA **62**, 550-557, 1969.

(10) Zubay, G., Morse, D.E., Schrenk, J.W. & Miller, J.H.M.: Detection and isolation of the repressor protein for the tryptophan operon. Proc.Natl.Acad.Sci. USA **69**, 1100-1103, 1972.

(11) Urm, E., Yang, H., Zubay, G., Kelker, N. & Maas, W.: *In vitro* repression of N-α-acetyl-L-ornithinase synthesis in *Escherichia coli*. Mol.Gen.Genet. **121**, 1-7, 1973.

(12) Martin Pollock recalls that during a small meeting of Pasteurians which he attended, someone asked Monod whether he knew that Pollock had been elected member of the Royal Society. "Oh yes: it was a mistake …", was his comment in front of Pollock. Pollock, M.: An exciting but exasperating personality. In: Origins of molecular biology. A tribute to Jacques Monod. Ed. by A. Lwoff & A. Ullmann. Academic Press, New York, San Fransisco, London, 61-73, 1979. Aaron Novick, who had been Szilard's collaborator, told me another story. When he arrived in 1954 at the Pasteur to work in Monod's group, he found in his hotel a note that he should present himself at the American Embassy. There, he should sign the loyalty oath, that the Rockefeller foundation, which was

paying his stipend, had forgotten to ask for. Novick was awake all night discussing his situation with his wife. He came to the conclusion that he would not sign but would instead return to an uncertain future in the US. Early in the morning he went to the lab to inform Monod about his decision. There he found Monod and Delbrück who burst out with laughter over their prank.

1.10 Putting the *lac* Genes on Phage φ80 or λ

A science like molecular biology advances with new methods and new ideas. Often a new idea needs a new method in order to be tested. If we look back, we notice that the study of the *lac* system got a tremendous boost from the synthesis and commercial availability of various substrates and inducers of the *lac* system. After the substrate o-nitrophenyl-β-D-galactoside (ONPG) became available, the *lac* system was the easiest system to test. The synthesis of isopropyl-1-thio-β-D-galactoside (IPTG) made possible the quantitative study of induction. The synthesis of 5-bromo-4-chloro-3-indolyl-β-D-galactoside (Xgal) allowed for the rapid recognition of colonies or tissues which contain β-galactosidase. Gene mapping and dominance studies required the existence of *Hfr* strains and *F'lac* episomes. It is significant how little time was wasted after their discovery. They were applied immediately in the *lac* system.

A similar situation characterized the analysis of the *lac* operator. The *lac* operator is by definition the DNA region to which Lac repressor binds. A simple calculation ($4^n = 4.7 \times 10^6$; $n \sim 11$) shows that any operator need not be larger than a dozen base pairs, as a sequence of eleven base pairs should occur once in the *E. coli* chromosome (4.7×10^6 BP) by chance. But how could one analyse such a small DNA region buried in so much other DNA? Here the λ *CI*/operator system has a clear advantage. λ DNA consists of 49,000 base pairs. Thus the λ operator, when present in λ phage DNA, is purified one hundredfold compared to its situation when present in *E. coli* DNA.

Phage λ inserts at a specific site of the *E. coli* chromosome, between the galactose and the biotin genes. Occasionally when it excises, it is imprecisely excised. Then it may incorporate some of the *E. coli* genes from one or the other side, sometimes some or all of the galactose genes, sometimes some or all of the biotin genes. Sometimes it does not lose any important genes in this process. Then the new phage can replicate. Sometimes it does lose one or several important genes. Then it can still replicate, but only in the presence of wildtype λ phage and its proteins. This never happens with the *lac* genes, as they are situated much too far away from the λ attachment site.

Three other phages have been described which are similar in sequence and structure to λ (*φ80, 434, HK022*). They attach and insert specifically in a similar manner, but elsewhere in the *E. coli* chromosome. All three of them insert far away from the *lac* genes. Now it happened that an American postdoc working in the Institut Pasteur, Jon Beckwith, observed that a *F'lac* episome had integrated into the *E. coli* chromosome at the site of the *TonB* gene, thereby making the bacteria resistant to phage *T1* (The *TonB* gene plays a role in the attachement of phage *T1* at the outer membrane of *E. coli*). Another American postdoc, Ethan Signer, recalled that the *TonB* gene was closely linked to the attachment site of phage φ*80* on the *E. coli* chromosome. He suggested that they should try to obtain φ*80(d)lac* phage from such bacteria (a phage characterized as *d,* where *d* stands for defective, needs the help of another phage to replicate). They did so and indeed obtained φ*80lac* phage (1)!

This trick could be generalized. Thus, in principle it became possible to put (almost) any gene on phage λ or φ*80* (2). Driving the technique to the utmost extreme, it was even possible to put the *lac* genes in one orientation on φ*80* and in the alternative on λ phages (3). If one denatured the two phage DNAs carrying the *lac* DNAs, and then renatured them after mixing them together, one occasionally obtained molecules where only the *lac* genes had formed double stranded DNA. The λ and φ*80* DNAs could not form double stranded DNA. They remained single stranded and could be degraded by a specific nuclease. Only *lac* DNA which was double stranded resisted degradation. One could see the double stranded *lac* DNA under the electron microscope (3). Thus for the first time a particular gene – the *lacZ* gene – was seen. The authors gave a press conference. They declared that they were frightened by their discovery (4). As a result of this discovery one author, James Shapiro, decided to leave science. He went to Cuba, became disillusioned and later returned to the United States, where he is now a professor at the University of Chicago.

The technique of "cloning DNA" on a phage like λ worked for *E. coli* genes (5). But it was laborious. When a few years later *in vitro* cloning was discovered this *in vivo* technique was largely abandoned. The new *in vitro* techniques were so much easier and general.

References

(1) Beckwith, J.R. & Signer, E.R.: Transposition of the *lac* region of *E. coli*. I. Inversion of the *lac* operon and transduction of *lac* by φ*80*. J.Mol.Biol. **19**, 254-265, 1966.

(2) Beckwith, J.R., Signer, E.R. & Epstein, W.: Transposition of the *lac* region. In: Cold Spring Harbor Symposia Quant.Biol. **31**, 393-401, 1966.

(3) Shapiro, J., MacHattie, L., Eron, L., Ihler, G., Ippen, K., Beckwith, J.R., Arditti, R., Reznikoff, W. & MacGillivray, R.: The isolation of pure *lac* operon DNA. Nature **224**, 768-774, 1969.

(4) Burch, S.: Scientists find the secret of human heredity – and it scares them. The frightening fact of life. Daily Mail, page 1, November 24, 1969. "Dr. Beckwith, 33, said: I fear the bad rather outweighs the good in this particular work. I feel that it is more frightening than hopeful."

(5) Silhavy, T.J., Berman, M.L. & Enquist, L.W.: Experiments with gene fusions. Cold Spring Harbor Laboratory, 1984.

1.11 Lac and λ *CI* Repressors Bind to their Operator DNAs

CI binds operator expt

Mark Ptashne isolated λ *CI* repressor by specifically labeling it with radioactive amino acids (1). This invited him to test whether λ *CI* repressor indeed binds to λ operator DNA, as the Jacob-Monod theory predicted. All he had to do was to see whether radioactive λ *CI* repressor (which presumably had a molecular weight less than 100,000 Daltons) sedimented together with λ DNA (which had a molecular weight of about 35,000,000 Daltons) in a glucose or glycerol gradient. (I recall here that glucose or glycerol gradients serve only to restrict possible uncontrolled movement of the buffer in the tube during centrifugation.) λ *vir* DNA, which replicates in the presence of λ *CI* repressor due to mutations in λ operator, was used as a control. The experiment worked spectacularly well: λ *CI* repressor sedimented together with wildtype λ DNA, but poorly with λ *vir* DNA (2,3).

When the experiment was done, the sequences of the two λ operator regions were unknown. Today we know that λ contains two operator regions and that each consists of three operators of slightly varying sequence and strength. λ *vir* mutants may contain one mutation in the left operator region (in O_L1) and two mutations in the right operator region (in O_R1 and O_R2) (4). In the presence of relatively low concentrations of λ *CI* repressor, two λ *CI* repressor dimers bind cooperatively to two neighbouring operators. This binding is reduced in the λ *vir* mutant DNA which carries mutations in the left *and* the right operator regions. I note here a curiosity: one may think that λ *CI* repressor-operator binding may have been subsequently analysed in detail with gel shift analysis, an easy method that has been available for almost ten years. Most curiously, this has never been done. Gel shifts with the λ system supposedly cannot be interpreted.

The λ experiment invited to perform the analogous experiment with Lac repressor and *φ80lac* phage DNA. Here Lac repressor had to be labelled. This was done by Walter Gilbert by growing *E. coli* carrying three sets of the *lac I⁺* gene in the presence of radioactive $^{35}SO_4^-$ and by purifying the labelled Lac repressor by ammonium sulfate precipitation and sephadex chromatography. The experiment worked very well (5). Radioactive Lac repressor sedimented with *φ80lac* DNA. The controls with two O^c mutations of various strength and with added inducer IPTG gave the expected results. Cosedimentation was reduced or absent.

These two experiments proved beyond a doubt that the central element of the Jacob-Monod theory was correct: repressor binds specifically to operator DNA. However, the tests were laborious and not suitable for routine analysis. A rapid test was introduced by Arthur Riggs and Suzanne Bourgeois (6). They recalled, but did not cite the fact that ribosomes bind very tightly to nitrocellulose (7). Lac repressor, like most proteins, does so as well. However, double stranded DNA does not bind to the filter. Thus, if one uses ^{32}P labelled φ*80lac* DNA and nitrocellulose filters, one observes the binding of the labelled DNA to the filter only in the presence of Lac repressor. DNA passes through the filter in the absence of Lac repressor. The test works nicely with Lac repressor but poorly with some other repressors such as Trp repressor (8). The test was used extensively for the study of the kinetics of *lac* repressor-operator interaction. I will return to this later. I merely mention here a curious phenomenon: phage DNA which carried *lac* DNA in which the *I* gene, the *lac* operator and a short region of the *lacZ* gene were deleted, still bound to Lac repressor (9,10). The first part of the *lacZ* gene contains another *lac* operator like region which binds with an efficiency of about ten percent of that of the normal *lac* operator. Its sequence resembles the sequence of *lac* operator. Another such sequence of worse quality (10) was found between the 3' end of the *I* gene and *lac* promoter. Both sequences seemed to have no function (11). So they were called *pseudooperators*.

have no f(x)

References

(1) Ptashne, M.: Isolation of the λ phage repressor. Proc.Natl.Acad.Sci. USA **57**, 306-313, 1967.

(2) Ptashne, M.: Specific binding of the λ phage repressor. Nature **214**, 232-234, 1967.

(3) Ptashne, M. & Hopkins, N.: The operators controlled by the λ phage repressor. Proc.Natl.Acad.Sci. USA **60**, 1282-1287, 1968.

(4) Flashman, S.M.: Mutational analysis of the operators of bacteriophage λ. Mol.Gen.Genet. **166**, 61-73, 1978.

(5) Gilbert, W. & Müller-Hill, B.: The *lac* operator is DNA. Proc.Natl.Acad.Sci. USA **58**, 2415-2421, 1967.

(6) Riggs, A.D. & Bourgeois, S.: On the assay, isolation and characterisation of the *lac* repressor. J.Mol.Biol. **34**, 361-368, 1968.

(7) Leder, P. & Nirenberg, M.: RNA codewords and protein synthesis II. Nucleotide sequence of a valine RNA codeword. Proc.Natl.Acad.Sci. USA **52**, 420-427, 1964.

(8) Klig, L.S., Crawford, I.P. & Yanofsky, C.: Analysis of *trp* repressor-operator interactions by filter-binding. Nucl.Acids Res. **15**, 5339-5351, 1987.

(9) Reznikoff, W.S., Winter, R.B. & Hurley, C.K.: The location of the repressor binding sites in the *lac* operon. Proc.Natl.Acad.Sci. **71**, 2314-2318, 1974.

(10) Gilbert, W., Majors, J. & Maxam, A.: Lactose operator sequences and the action of Lac repressor. In: Symposium on Protein-Ligand Interactions. Ed. by H. Sund & G. Blauer. Walter de Gruyter, Berlin. 193-206, 1975.

(11) Pfahl, M., Gulde, V. & Bourgeois, S.: "Second" and "third operator" of the *lac* operon: an investigation of their role in the regulatory mechanism. J.Mol.Biol. **127**, 339-344, 1979.

1.12 Making Grams of Lac Repressor to Determine its Sequence

ppl need to make large amounts of LacI

Lac and λ *CI* repressor were isolated. But they were present in such small amounts that one could never produce enough material for use in any biochemical analysis. Lac repressor became the first repressor for which a strategy was devised to make gram amounts (1). In fact, for about ten years it remained the only system which provided large amounts of repressor. This situation changed only after the advent of *in vitro* DNA manipulation and cloning. Yet the strategy used was straight forward.

1. A mutant was isolated which overproduced Lac repressor (1). As starting material a temperature sensitive I^{TS} mutant (2,3) was used. Mutagenized bacteria were grown at the highest possible temperature (44°C) in the presence of o-nitrophenyl-1-thio-β-D-galactoside, which is accumulated by Lac permease and which is toxic at high concentrations. Thus *lac* constitutive mutants are selected against at high temperature, where almost all of the Lac repressor is inactive. A substantial increase in mRNA synthesis would lead to the formation of sufficient amounts of active Lac repressor to repress the synthesis of *lac* operator-dependent β-galactosidase. Therefore, the bacteria were plated at 42°. Colonies which had repressed levels of β-galactosidase were purified. Two colonies produced about tenfold more repressor than wildtype I^+ at low temperature (30°C). They were called I^Q (Q stands for quantity). Later the TS mutation was crossed out by Jeffrey Miller, so that wildtype and not TS repressor is overproduced.

2. The number of *I* genes was increased. For this, one I^{QTS} mutation was crossed into a hybrid λ *h80dlac* phage (it was part $\phi 80$, part λ) which was inserted into the *E. coli* chromosome. The phage carried two mutations which made its handling easy. First, it carried a mutation in the λ *CI* repressor gene producing temperature sensitive λ *CI* repressor (CI_{857}). Thus it could be induced by a temperature shift from 32° to 42°. Second, it carried a mutation (*t68*) which made it impossible for the phage to lyse the bacteria. Thus, after heat induction, more and more copies of λ and *lacI* DNA are produced. The bacteria die but do not lyse. This trick increases the number of *I* genes so efficiently that it overproduces Lac repressor about thirtyfold (1).

When combined with the I^Q mutation, the amount of repressor is increased about three hundred-fold. Subsequently an improved technique for isolating I^Q mutations became available (4). The best of the I^Q mutants overproduces Lac repressor about one hundred-fold. It was called I^{Q1}. Coupled with the phage, it overproduces Lac repressor several thousand-fold (5). As soon as DNA sequencing became possible, the I^Q and the I^{Q1} mutant were sequenced. Both contained mutations in the promoter. In the I^Q mutation a bad -35 site was optimized (6), in I^{Q1} a deletion of 15bp created an optimal -35 site (7).

Ammonium sulfate precipitation and phosphocellulose chromatography are sufficient to produce pure Lac repressor from lysates of these bacteria (8). Each monomer binds one molecule of IPTG under standard conditions (8). (There is, however, the report of Oshima *et al.* (9), never repeated by others, that each tetramer binds only 2.5 molecules of IPTG at low temperature and high pH.) So, more than ten grams of Lac repressor was produced from a few kilograms of *E. coli* cells. The bulk of the material was used to determine the amino acid sequence of Lac repressor (10). It was a tetramer in which each monomer appeared to be composed of 347 residues. The DNA sequence of the *lacI* gene, obtained five years later (11) by a graduate student of Walter Gilbert, indicated that thirteen amino acids had been missed (11,12): In fact, Lac repressor contains 360 residues. At the time the sequence of Lac repressor was rather disappointing in that it did not show any peculiarities that could be correlated with its activities. A Lys-Arg-Lys sequence close to the C-terminus seemed histone-like and therefore possibly involved in DNA backbone binding (10) (It seems not to be). But no more could be seen.

The sequence of Lac repressor was the first and for five years the only sequence of a repressor which was determined entirely by protein sequence analysis. Then the protein sequence of λ *CI* repressor was determined (13). All subsequent sequences of repressors have been determined fully only as DNA sequences. It is noteworthy that the technique of overproducing Lac repressor was not used for any other system. It could not work in the λ system due to the autoregulation of *CI* expression. When the attempt was finally made to overproduce λ *CI* repressor by *in vivo* gene fusion (14), the *in vitro* techniques seemed so much more interesting and promising (15) that the old fashioned construct was almost never used (16) and thus immediately forgotten.

In the early seventies enough pure Lac repressor was made in Cologne so that several hundred milligrams were given away, mainly to X-ray crystallographers who tried to crystallize it. Yet this was all done in vain. At the time no one succeeded in obtaining crystals suitable for X-ray analysis . Here, luck was not on the side of Lac repressor.

References

(1) Müller-Hill, B., Crapo, L. & Gilbert, W.: Mutants that make more *lac* repressor. Proc.Natl.Acad.Sci. USA **59**, 1259-1264, 1968.

(2) Horiuchi, T. & Novick, A.: Studies of a thermolabile repressor. Biochem.Biophys.Acta **108**, 687-696, 1965.

(3) Sadler, J.R. & Novick, A.: The properties of repressor and the kinetics of its action. J.Mol.Biol. **12**, 304-327, 1965.

(4) Starting material was a fusion between a wildtype *lac I⁺* gene and a *lac Z⁺* gene which produces active β-galactosidase in amounts which are insufficient for growth on lactose. *E. coli* carrying this fusion is plated on lactose. Most *lac⁺* revertants carry promoter-up mutations in the *lac I* gene. I used this method extensively, but never published it.

(5) Müller-Hill, B.: Lac repressor and *lac* operator. Progr.Biophys.Molec.Biol. **30**, 227-252, 1975.

(6) Calos. M.P.: DNA sequence for a low-level promoter of the *lac* repressor gene and an "up" promoter mutation. Nature **274**, 762-765, 1978.

(7) Calos, M.P. & Miller, J.H.: The DNA sequence change resulting from the I^{Q1} mutation which greatly increases promoter strength. Mol.Gen.Genet. **183**, 559-560, 1981.

(8) Müller-Hill, B., Beyreuther, K. & Gilbert, W.: Lac repressor from *Escherichia coli*. Methods in Enzymology **21**, Part D, 483-487, 1971.

(9) Oshima, Y., Mizokoshi, T, & Horiuchi, T.: Binding of an inducer to the *lac* repressor. J.Mol.Biol. **89**, 127-136, 1974.

(10) Beyreuther, K., Adler, K., Geisler, N. & Klemm, A.: The amino-acid sequence of *lac* repressor. Proc.Natl.Acad.Sci. USA **70**, 3576-3580, 1973.

(11) Farabough, P.J.: Sequence of the *lacI* gene. Nature **274**, 765-769, 1978.

(12) Beyreuther, K.: Revised sequence of the *lac* repressor. Nature **274**, 767, 1978.

(13) Sauer, R.T. & Anderegg, R.: Primary structure of the λ *CI* repressor. Biochemistry **17**, 1092-1100, 1978.

(14) Gronenborn, B.: Overproduction of phage λ *CI* repressor under control of the *lac* promoter of *Escherichia coli*. Mol.Gen.Genet. **148**, 243-250, 1976.

(15) Backman, K.M., Ptashne, M. & Gilbert, W.: Construction of plasmids carrying the *CI* gene of bacteriophage. Proc.Natl.Acad.Sci. USA **73**, 4174-4178, 1976.

(16) Beyreuther, K. & Gronenborn, B.: N-Terminal sequence of phage λ repressor. Mol.Gen.Genet. **147**, 115-117, 1976.

1.13 Mutations in *lacI* Suggest a Modular Structure of Repressor

The operator had been defined as the DNA region which interacts with repressor and where operator constitutive (O^c) mutations are found. Now, when it became clear that transcription of mRNA starts at particular points on the DNA, it made sense to postulate DNA-regions, the promoters, which interact with RNA polymerase (1). Mutations in this region (P^- or P^+) should have lower or higher levels of expression of all translation products of the particular mRNA. Such *lacP*$^-$ mutants were found (1). Here, the expression of β-galactosidase, permease and transacetylase was lowered by factors from ten to fifty. In contrast to the O^o mutants which I discussed earlier, the expression remained sizable. It now became a challenge to map these P^- mutations with respect to mutations in the operator, the *I* and the *Z* genes.

In 1965, the order of the genetic elements of the *lac* system was wrongly determined to be *IOPZY* (instead of *IPOZY*) by Jacob and Monod (2). The three factor crosses which these authors used are not reliable when the mutations are very close to one another. Thus the data were wrongly interpreted. However, genetic ordering of even very close markers can be done reliably by deletion mapping. This was done in the laboratory of Jon Beckwith (3). They showed that the correct order of the genetic elements is *IPOZY*. I should add that at the time the direction of transcription of the *lacI* gene was not known. The position of the promoter of the *I* gene was determined in 1968: *PIPOZY* (4).

The wrong order *IOPZ* invited a search for deletions which presumably fuse the inactivated *I* gene with the *lac* operator. They should be I^- and O^c. Such mutants were sought by Julian Davies (5). Surprisingly, a few dominant constitutive I^- mutations were found, in spite of the fact that during the work the order of the genetic elements was shown to be wrong. When these mutants were tested for their dominance it turned out that they were *trans*-dominant (I^{-d}) (d stands for dominant) and not *cis*-dominant (5)! It was proposed that they were defective in the domain of Lac repressor which binds specifically to *lac* operator (5,6).

To analyse this type of mutant in more detail, first 60 and later 187 such I^{-d} mutants were isolated (7,8). Most of them bound the inducer IPTG normally *in vitro*. Thus, their inducer binding domain was intact. Their (negative) transdominance was explained by the proposal that they formed mixed tetramers with the I^+ wildtype subunits of Lac repressor (6-8). If the negative dominant mutation is driven by an I^Q promoter, the number of repressor molecules which have four intact subunits should decrease drastically, *i.e.* it should become less then one per cell. One then has to argue that all four subunits of Lac repressor are necessary for operator binding, *i.e.* effective repression.

At that time Jeffrey Miller (see next chapter!) began to collect *TonB-I* deletions, which enter into the *I* gene from the promoter (4). By the early seventies 39 such deletions were isolated. All 247 I^{-d} mutations were mapped against this set of 39 *TonB-I* deletions (7,8). The mapping indicated that almost all strong I^{-d} mutations are located within the first seven (of 39) deletion groups, *i.e.* they all map at the promoter proximal end of the *I* gene. If one assumes that the deletions occur randomly and that repressor consists of 340 residues, all mutations map within the first fifty codons of the *I* gene. This conclusion was supported by Klaus Weber, Jeffrey Miller and their collaborators, who sequenced two N-terminal repressor mutants they had mapped with the same set of deletions (9). Later the protein sequence of 29 I^{-d} mutants confirmed the genetic allocation fully (10). Thus it was proposed that an N-terminal protrusion of Lac repressor, about fifty residues long, is responsible for specific binding to *lac* operator (7).

When Lac repressor is treated with trypsin, it is cut first only at Arg51 or Lys59. The resulting N-terminal fragments can be isolated (11). They were called headpieces. When the headpieces themselves were tested for the capacity to bind specifically to the *lac* operator they were indeed found to be capable of doing so (12). For this experiment, milligram amounts of headpieces had to be used to protect *lac* operator: the headpieces lack the capability of aggregating into suitably spaced dimers.

But how does this short N-terminal piece of Lac repressor recognize DNA? It had been shown (13) that an alpha helix fits sterically into the major groove of DNA. A plausible alpha helical region seemed to begin at residue 17 of Lac repressor. So I predicted (7) that a helix beginning at tyrosine 17 recognizes the major groove of the Lac operator. That this was a likely possibility was shown ten years later, when the Helix-Turn-Helix (HTH) motif of Lac repressor was demonstrated to be similar in sequence to the HTH motifs of λ *CI* repressor and λ *cro*, for the latter of which a crystal structure existed (14). Formal proof that the prediction was correct came 15 years later through NMR analysis (15).

At the time these experiments were done, it was not known that proteins generally have a modular structure. One domain of a protein may have one activity, the next domain another activity and so on. Each domain corresponds to a linear part of its gene. In the case of Lac repressor, the domain specifying inducer binding, the core, was mapped using I^s mutations. This was done with 90 I^s mutants and indeed they mapped in clusters in the middle of the *I* gene and at the distal end of the I^{-d} cluster (8). Thus the operator and inducer binding domains occupy separate parts of Lac repressor. Similar mutants were found in the *CI* gene of λ. Their map positions indicated that the same was true for λ *CI* repressor. The domain of λ *CI* repressor which determines operator binding is positioned at the extreme N-terminus; the position of the region important for UV induction is in the core (16). Once more Lac and λ *CI* repressor looked rather similar!

Finally it was asked whether there are mutations in which Lac repressor is unable to form a tetramer? Their phenotype is easily defined. They should be recessive-constitutive and they should bind *in vitro* inducer IPTG. A search of one hundred recessive constitutives yielded one such mutant (17). It mapped near codon 220. A search of seven hundred constitutives yielded eight such mutants (18). They mapped around codon 224 and between codons 273 and 283. As predicted, such mutant repressor was not tetrameric but consisted of a mixture of monomers and dimers.

References

(1) Jacob, F., Ullmann, A. & Monod, J.: Le promoteur, élement génétique nécessaire à l'expression d'un operon. Comptes Rendus Acad.Sci. **258**, 3125-3128, 1964.

(2) Jacob, F. & Monod, J.: Genetic mapping of the elements of the lactose region. Biochem.Biophys.Res. Communications **18**, 693-701, 1965.

(3) Miller, J.H., Ippen, K., Scaife, J.G. & Beckwith, J.R.: The promoter-operator region of the *lac* operon of *Escherichia coli*. J.Mol.Biol. **38**, 413-420, 1968.

(4) Miller, J.H., Beckwith, J.R. & Müller-Hill, B.: Direction of transcription of a regulatory gene in *E. coli*. Nature **220**, 1287-1290, 1968.

(5) Davies, J. & Jacob, F.: Genetic mapping of the regulator and operator genes of the *lac* operon. J.Mol.Biol. **36**, 413-417, 1968.

(6) Müller-Hill, B., Crapo, L. & Gilbert, W.: Mutants that make more *lac* repressor. Proc.Natl.Acad.Sci. USA **59**, 1259-1264, 1968.

(7) Adler, K., Beyreuther, K., Fanning, E., Geisler, N., Gronenborn, B., Klemm, A., Müller-Hill, B., Pfahl, M. & Schmitz, A.: How *lac* repressor binds to DNA. Nature **237**, 322-327, 1972.

(8) Pfahl, M., Stockter, C. & Gronenborn, B.: Genetic analysis of the active sites in *lac* repressor. Genetics **76**, 669-679, 1974.

 (9) Weber, K., Platt, T., Ganem, D. & Miller, J.H.: Altered sequences changing the operator-binding properties of the *lac* repressor: Colinearity of the repressor protein with the *I*-gene map. Proc.Natl.Acad.Sci. USA **69**, 3624-3628, 1972.

(10) Schlotmann, M. & Beyreuther, K.: Degradation of the DNA-binding domain of wild-type and *I⁻ᵈ lac* repressor in *Escherichia coli*. Eur.J.Biochem. **95**, 39-49, 1979.

(11) Geisler, N. & Weber, K.: Isolation of the N-terminal fragment of lactose repressor necessary for DNA binding. Biochemistry **16**, 938-943, 1977.

(12) Ogata, R.T. & Gilbert, W.: An amino-terminal fragment of *lac* repressor binds specifically to *lac* operator. Proc.Natl.Acad.Sci. USA **75**, 5851-5854, 1978.

(13) Sung, M.T. & Dixon, G.H.: Modification of histones during spermatogenesis in trout: A molecular mechanism for altering histone binding to DNA. Proc.Natl.Acad.Sci. USA, **67**, 1616-1623, 1970.

(14) Matthews, B.W., Ohlendorf, D.H., Anderson, W.F. & Takeda, Y.: Structure of the DNA binding region of *lac* repressor inferred from its homology with *cro* repressor. Proc.Natl.Acad.Sci. USA. **79**, 1428-1432, 1982.

(15) Boelens, R., Scheek, R.M., van Boom, J.H. & Kaptein, R.: Complex of *lac* repressor headpiece with a 14 base-pair *lac* operator fragment studied by two-dimensional nuclear magnetic resonance. J.Mol.Biol. **193**, 213-216, 1987.

(16) Müller-Hill, B., Gronenborn, B., Kania, J., Schlotmann, M. & Beyreuther, K.: Similarities between Lac repressor and λ *CI* repressor. In: Nucleic acid-protein recognition. Ed. by J. Vogel. Academic Press, New York, San Francisco, London. 219-236, 1977.

(17) Müller-Hill, B.: *Lac* repressor and *lac* operator. Progr.Biophys.Mol.Biol. **30**, 227-252, 1975.

(18) Schmitz, A., Schmeissner, U. & Miller, J.H.: Mutations affecting the quarternary structure of the *lac* repressor. J.Biol.Chem. **251**, 3359-3366, 1976.

1.14 Miller's Analysis of the *I* Gene: Climbing Mount Everest

The genetic analysis of the modular structure of Lac repressor (described in the preceeding chapter) was advanced at its time. It looks pedestrian when compared to the project started by Jeffrey Miller at the same time. He isolated 5500 *I⁻* nonsense mutants, more than 2000 *I⁻* missense mutants and maped them all with the help of more than 400 *TonB-I* deletions. He presented his work at the Cold Spring Harbor conference in 1977 and wrote an excellent review for the book resulting from this conference (1).

His original goal was to produce a genetic fine structure map of the *I* gene (2,3) which would correlate with the known protein sequence of Lac repressor (4). The moment the sequence of the *lacI* gene became known, he correlated it with his genetic map (5). How was the genetic map produced? Miller isolated more than 400 *TonB-I* deletions. These deletions are resistant against phage φ80 *vir* and colicin V,B *and* they are *lac* constitutive. The *TonB* gene and the *I* gene are direct

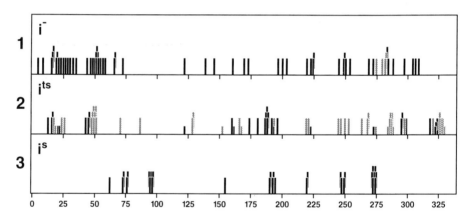

Fig. 5.: The distribution of mutational sites in the *lacI* gene. The top panel shows the sites resulting in the constitutive *I⁻* phenotype, the second panel those resulting in the heat sensitive, constitutive *Iᵗˢ* phenotype, the third panel in the noninducible *Iˢ* phenotype. The horizontal scale gives the position of the corresponding residue of Lac repressor. The data were produced analysing over 1000 independent mutations. Open bars in the first panel: aggregation deficient mutants; second panel: weaker phenotypes. Adapted from (7), with permission.

neighbours in the construct Miller used. Thus, a mutant which is *TonB* resistant and *lac* constitutive has to be a deletion which extends from the *TonB* gene into the *lacI* gene. Miller made use of selection protocolls which allowed the isolation of rare I^+ recombinants from frequent I^- or I^s point mutants. The method he used was so sensitive that it allowed the isolation of I^+ recombinants when the recombination had to occur within very few base pairs.

Miller accurately defined more than 90 *I*-deletions with more than 70 I^- amber mutants (6) and a large number of *I* missense mutants (7). Every I^- amber mutant was suppressed by five different amber suppressors (1,6), thus 70 residues of Lac repressor could be replaced by serine, glutamine, tyrosine, leucine or lysine. It was tested whether these mutant repressors behaved like wildtype (I^+), were constitutive (I^-), temperature sensitive (I^{TS}) or uninducible (I^s). The results showed clearly that specific mutations are clustered in specific locations (Fig. 5).

References

(1) Miller. J.H.: The *lacI* gene: Its role in *lac* operon control and its use as a genetic system. In: The operon. Ed. by J.H. Miller & W.S. Reznikoff. Cold Spring Harbor Laboratory, Cold Spring Harbor, N.Y., 31-88, 1978.

(2) Miller, J.H., Ganem, D., Lu, P. & Schmitz, A.: Genetic studies of the *lac* repressor, I. Correlation of mutational sites with specific amino acid residues: construction of a colinear gene-protein map. J.Mol.Biol. **109**, 275-302, 1977.

(3) Schmeissner, U., Ganem, D. & Miller, J.H.: Genetic studies of the *lac* repressor, II. Fine structure deletion map of the *lacI* gene, and its correlation with the physical map. J.Mol.Biol. **109**, 303-326, 1977.

(4) Beyreuther, K., Adler, K., Geisler, N. & Klemm, A.: The amino-acid sequence of *lac* repressor. Proc.Natl.Acad.Sci. USA **70**, 3576-3580, 1973.

(5) Miller, J.H., Coulondre, C. & Farabough, P.J.: Correlation of nonsense sites in the *lacI* gene with specific codons on the nucleotide sequence. Nature **274**, 770-775, 1978.

(6) Miller, J.H., Coulondre, C., Hofer, M., Schmeissner, U., Sommer, H., Schmitz, A. & Lu, P.: Genetic studies of the *lac* repressor, IX. Generation of altered proteins by the suppression of nonsense mutations. J.Mol.Biol. **131**, 191-222, 1979.

(7) Miller, J.H. & Schmeissner, U.: Genetic studies of the *lac* repressor, X. Analysis of missense mutations in the *lacI* gene. J.Mol.Biol. **131**, 223-248, 1979.

1.15 Positive Control through the CAP Protein

What had happened to diauxie, the phenomenon Monod had described in his thesis, and which had stimulated his interest in the *lac* system of *E. coli*? Monod had turned away from it. It seemed too complicated. Nontheless several other people continued to work on the problem, the glucose effect, as it was later called. On close analysis, several effects could be discerned. Glucose competes with some sugars for entry into *E. coli*. But glucose also has other fast and slow effects. It was a mess. Boris Magasanik, another refugee from Vienna, became the main investigator. He did not like the term glucose effect. Instead he called it catabolite repression (1).

The field changed radically in 1965 when two biochemists, who had nothing to do with the *lac* or any other such system, found that the level of cyclic AMP (cAMP) increased strongly in glucose starved *E. coli* cells (2). The authors of this paper saw no connection between their effect and diauxie, the glucose effect or catabolite repression. But Ira Pastan at the National Institutes of Health (NIH) (3) and Agnes Ullmann at the Pasteur (4) got the message. Independently they tested whether cAMP could antagonize the repression of β-galactosidase synthesis by glucose. Their experiments were quickly done and fully convincing: cAMP *did antagonize glucose repression* (3,4). This was the signal for Ira Pastan and Jon Beckwith and their groups to look for mutants which knock out the various parts of this regulatory system.

- Pastan and his collaborators looked for mutants in which the enzyme which produced cAMP was destroyed. In such mutants neither the *lac* nor any other catabolite sensitive system (for example the *ara* and the *mal* systems) should be fully expressed. One would expect these mutants to be pleiotropic *lac*, *ara* and *mal* negative. They should be phenotypically rescued by the addition of cAMP. Indeed such mutants were found on suitable indicator plates. Quantitative analysis indicated that the expression of β-galactosidase was reduced about fifty fold.
- Both groups isolated mutations which inactivated the protein which interacts with cAMP and which activates *lac* and various other promoters in the presence of cAMP. These mutants were not phenotypically rescued by addition of cAMP (6,7). The product of this gene was called the CAP or CRP protein. When the CAP protein was inactive, the synthesis of β-galactosidase was reduced by a factor of fifty.

- The groups of Beckwith and Magasanik isolated mutations which abolish the DNA site to which CAP protein bound. The mutants mapped close to the *lac* promoter. Like ordinary promoter mutants, these mutants showed a ten to twenty-fold decrease in β-galactosidase production. But where mutants in the promoter itself were still fully dependent on the presence of CAP protein and cAMP in their expression, these mutants were much less sensitive to the presence or absence of CAP protein (8). When the target for CAP protein is completely abolished, dependence on CAP protein is absent. On the other hand, when the promoter proper is damaged its dependence on the presence of CAP should not have changed at all.

  Genuine promoter mutants and CAP target mutants clustered at different map positions. This was shown unambiguously by deletion mapping (8-10). The order of the genetic elements is *-I-CAP-P-O-Z-*, where the CAP site is very close to the *lac* promoter.

- Finally *lac*+ revertants were isolated from *lac*− mutations in the CAP target site (11). In these mutations the *lac* promoter had essentially become independent of CAP. They no longer needed the CAP protein to function optimally. Later one such mutant, called UV5, was used for many purposes.

The CAP protein was purified (6,12,13). It bound strongly to phosphocellulose, like all DNA binding proteins. It was used *in vitro* to demonstrate its binding to DNA containing the *lac* promoter and to activate transcription from the *lac* promoter (6,12). These and other experiments (14) provided the pieces for the explanation of the phenomenon of diauxie. But how glucose influences the cAMP and CAP levels – this was quite another question. Monod was profoundly interested in solving it. At the time, in 1976, all odds were against him. He was *directeur* of the Institut Pasteur, and he was severely ill. Yet together with Agnes Ullmann, he managed to produce his last paper on a factor of low molecular weight which drastically increases catabolite repression and which cannot be counteracted by cAMP (15). What is the chemical nature of this factor? Possibly we may never know. It has yet to be elucidated. The Institut Pasteur shifted its resources back to applied research. In this context the catabolite modulator factor was of no interest. Research on this topic could not be continued. And to this day nobody else has solved the problem.

References

(1) Magasanik, B.: Catabolite repression. Cold Spring Harbor Symp.Quant.Biol. **26**, 249-356, 1961.

(2) Makman, R.S. & Sutherland, E.W.: Adenosine 3', 5'-phosphate in *Escherichia coli*. J.Biol.Chem. **240**, 1309-1314, 1965.

(3) Perlman, R. & Pastan, I.: Cyclic 3'5'-AMP: stimulation of β-galactosidase and tryptophanase induction in *E. coli*. Biochem.Biophys.Res.Commun. **30**, 656-664, 1968. Perlman and Pastan were lucky: They used a strain of *E. coli* which lacks cAMP phophodiesterase when grown in mineral salts medium. So they could use a rather low concentration of cAMP and see an effect.

(4) Ullmann, A. & Monod, J.: Cyclic AMP as an antagonist of catabolite repression in *Escherichia coli*. FEBS Letters **2**, 714-717, 1968.

(5) Perlman, R.L. & Pastan, I.: Pleiotropic deficiency of carbohydrate utilization in an adenyl cyclase deficient mutant of *Escherichia coli*. Biochem.Biophy.Res.Commun. **37**, 151-157, 1969.

(6) Emmer, M., de Crombrugghe, B., Pastan, I. & Perlman, R.L.: Cyclic AMP receptor protein of *E. coli*. Proc.Natl.Acad.Sci. USA **66**, 480-487, 1970.

(7) Schwartz, D. & Beckwith, J.R.: Mutants missing a factor necessary for the expression of catabolite-sensitive operons in *E. coli*. In: The Lactose Operon. Ed. by J.R. Beckwith & D. Zipser. Cold Spring Harbor Laboratory, Cold Spring Harbor, New York. 417-422, 1970.

(8) Silverstone, A., Magasanik, B., Reznikoff, W.S., Miller, J.H. & Beckwith, J.R.: Catabolite sensitive site of the lac operon. Nature **221**, 1012-1014, 1969.

(9) Beckwith, J.R., Grodzicker, T. & Arditti, R.: Evidence for two sites in the *lac* promoter region. J.Mol.Biol. **69**, 155-160, 1972.

(10) Hopkins, J.D.: A new class of promoter mutations in the lactose operon of *Escherichia coli*. J.Mol.Biol. **87**, 715-724, 1974.

(11) Silverstone, A.E., Arditti, R.R. & Magasanik, B.: Catabolite insensitive revertants of *lac* promoter mutations. Proc.Natl.Acad.Sci. USA **66**, 773-779, 1970.

(12) Zubay, G., Schwartz, D. & Beckwith, J.R.: Mechanism of activation of catabolite sensitive genes: a positive control system. Proc.Natl.Acad.Sci. USA **66**, 104-110, 1970.

(13) Riggs, D., Reiness, G. & Zubay, G.: Purification and DNA-binding properties of the catabolite gene activator protein. Proc.Natl.Acad.Sci. USA **68**, 1222-1225, 1971.

(14) Perlman, R.L., de Crombrugghe, B. & Pastan, I.: cyclic AMP regulates catabolite transient repression in *E. coli*. Nature **223**, 810-812, 1969.

(15) Ullmann, A., Tillier, F. & Monod, J.: Catabolite modulator factor: a possible mediator of catabolite repression in bacteria. Proc.Natl.Acad.Sci. USA **73**, 3476-3479, 1976.

1.16 Isolation and Sequence of *lac* Operator

By 1968 it was clear how the amino acid sequence of Lac repressor could be determined. It was just a time consuming procedure. In contrast to Lac repressor *lac* operator which was much smaller was much more difficult to attack. It was embedded in the midst of the 50,000 basepairs of λ DNA. No method existed to sequence DNA. How could one go about doing it?

Walter Gilbert decided to protect *lac* operator with Lac repressor and then to destroy all the remaining DNA with DNase. The DNA left should be *lac* operator. To label λ DNA, the bacteria were grown in 10ml medium containing $^{32}PO_4^-$. Huge amounts of label had to be used, since most of the label was lost in the medium and in the *E. coli* DNA. Gilbert used 100 mCurie (*i.e.* 2×10^{12} counts/minute) for each experiment. Since the half-life of ^{32}P is about two weeks, the experiment had to be repeated every second week. With an immense effort, it was shown that about 20 to 25 basepairs were protected by Lac repressor. But the protected material, the operator, was unsuitable for any sequence determination.

To determine the operator sequence, its DNA had to be transcribed into RNA. RNA could be sequenced. For this purpose much more template was needed. So 300 mg of phage λ was isolated from a culture grown in a 100 litre fermenter. 25 mg of Lac repressor were added to the DNA for operator protection. So, a suitable amount of *lac* operator was isolated which could be used for transcription. If RNA polymerase and only nucleotide triphosphates were added, the reaction could start anywhere. It was impossible to deduce a sequence of the products of such reactions. Thus, selected primers had to be used. In the end a 24 basepair sequence was determined (1):

```
        10 9 8 7 6 5 4 3 2 1 0 1 2 3 4 5 6 7 8 9 10
        <----------     <-  <-   ->  ->  ----------->
5'  T G G A A T T G T G A G C G G A T A A C A A T T 3'
3'  A C C T T A A C A C A C G C C T A T T G T T A A 5'
```

It can be seen that the DNA sequence is partially dyad symmetric. The subunits of a suitable protein dimer can interact with each half-site.

While this most difficult experiment was performed by Walter Gilbert and his student Allan Maxam, an easier route opened up. *lac* promoter mutants were isolated which acted independently of CAP. They were transcribed specifically *in vitro* just by RNA polymerase. Thus, Nancy Maizels, a student of Gilbert, purified 100 mg of λ *dlac* DNA and sheared it in a sonifier down to a size of approximally

1,000 basepairs (2). She then purified the fragments containing *lac* operator with Lac repressor on nitrocellulose filters. When these fragments were transcribed *in vitro* with suitable nucleotide triphosphates, RNA was obtained which could be sequenced by the classical RNA sequence method. It turned out that the m-RNA began with A10 of *lac* operator! The sequence of β-galactosidase began at basepair 42! Thus, it was proven that *lac* operator was located downstream of *lac* promoter.

Nancy Maizels then proceeded to isolate in this λ construct *lac O^c* mutants carrying the CAP independent UV5 promoter (3). The sequences of these *lac* operator mutants indeed indicated that basepairs in the DNA region protected by Lac repressor were altered.

The reader may wonder how it happened that a graduate student – Nancy Maizels – published her most important paper under her own name? Why was Gilbert, who suggested the experiment and who had helped her daily during the course of her work, not a co-author? This practice, that the superviser of a graduate student or a postdoctoral fellow only becomes co-author if she or he participates substantially with her or his own hands in the work, derived from James Watson, who had introduced it in his lab in the fifties; before it was quite general practise in genetics. Gilbert had become Watson's partner and continued this practice. For the students it is indeed optimal. For seven years I tried it myself in my own lab but in a world which does not honour it, it is a doomed enterprise. I capitulated earlier. Gilbert gave up later. These days one is co-author if one obtained the grant money for the research. And, to get grant money, one is judged by the papers where one is co-author.

References

(1) Gilbert, W. & Maxam, A.: The nucleotide sequence of the *lac* operator. Proc.Natl.Acad.Sci. USA **70**, 3581-3584, 1973.

(2) Maizels, N.M.: The nucleotide sequence of the lactose messenger ribonucleic acid transcribed from the UV5 promoter mutant of *Escherichia coli*. Proc.Natl.Acad.Sci. USA **70**, 3585-3589, 1973.

(3) Gilbert, W., Maizels, N.M. & Maxam, A.: Sequences of controlling regions of the lactose operon. Cold Spring Harbor Symposium **38**, 845-856, 1973.

1.17 Chemical DNA Sequencing

When Walter Gilbert and his technician Allan Maxam set out to determine the sequence of the *lac* operator, no method existed for sequencing DNA. They were able to get the *lac* operator RNA sequence in 1973. It helped them that one of Gilbert's graduate students, Nancy Maizels, sequenced with conventional methods a *lac* RNA transcript which contained the *lac* operator (see section 1.16.). It is characteristic of Gilbert that he would not let the situation stand as it was. He wanted to develop a *general* method for sequencing DNA.

The idea for a general method was straightforward, but its realisation entailed a major piece of chemical work (1). And I remind the reader, Gilbert was a theoretical physicist and not a chemist by training. It was his intelligence and courage combined with the hard work of his technician and later graduate student Maxam which enabled them to solve the problem. A suitable restriction fragment has to be labelled radioactively only at one end. Four samples of the labelled fragment are then analysed. Two are treated under identical conditions with dimethylsulfate, two others are treated with hydrazine under different conditions. The general chemistry of these reactions had already been worked out by others. Dimethylsulfate methylates the nitrogen in position 7 (7N) of guanine (G) and the nitrogen in position 3 (3N) of adenine (see Fig. 6). Hydrazine reacts in one sample both with cytosine (C) and with thymidine (T). High salt concentration of KC1 inhibits the reaction with T but allows the reaction with C in the other sample. The reaction conditions are so arranged that about one base is substituted in each DNA molecule. The substituted G's and A's are removed by heating, the substituted C's and T's by treatment with piperidin. The phosphate bonds with the free sugars are then hydrolysed with a solution of NaOH. The DNA backbones are hydrolyzed preferentially at the sugars which lack the base. The four reaction mixtures are then put on four tracks of an acrylamide gel which separates DNA strands according to their size. The sequence can be read directly from the gel (1).

The Maxam-Gilbert method is the most amazing chemical method conceived and executed by non-chemists: Walter Gilbert carried out the first experiments in 1974. Andrej Mirzabekov, a visitor from Moscow in the lab suggested treating a *lac O* containing DNA fragment with dimethylsulfate in the presence of Lac repressor. Gilbert performed this methylation protection experiment with a DNA restriction fragment carrying *lac* 01 in the presence and absence of Lac repressor under conditions where only about one G or A is methylated (2; see also

Major groove

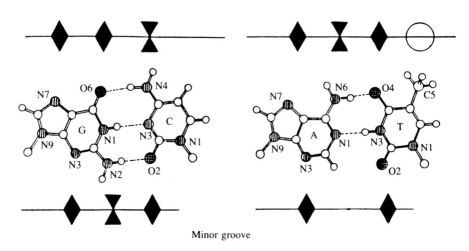

Minor groove

Fig. 6.:   The hydrogen-bond donors and acceptors presented by Watson-Crick base pairs to the major groove and the minor groove. The symbols for hydrogen bond donors (✕) and acceptors (◆) show a varied pattern presented by the base pairs to the major and an poor information array in the minor groove. While it is possible to distinguish among AT, TA, GC and CG in the major groove, functional groups in the minor groove allow only easy discrimination between AT and GC containing base pairs. O , methyl group. ● bond to deoxyribose. Fig. 2 of (11), with permission.

section 1.16.). Suddenly it became clear that one could define the positions of G's and A's in the sequence with this experiment. If a similar reaction for the C's and T's could be worked out one could read the sequence on the gels. But was Gilbert the first to think along these lines? One may ask, who first had the idea? And here comes a surprise. The idea of the Maxam-Gilbert method was published in 1973 by a Russian chemist, Eugene Sverdlov, in a western journal (3). Sverdlov and his collaborators had used a synthetic tetranucleotide to demonstrate that they could identify the positions of purines with this method. In their paper they list the following steps: 1. labelling of the oligonucleotide at one end. 2. specific modification of bases (purines). 3. specific cleavage of the sugar phosphate bonds carrying the modified bases. 4. Separation of the reaction products by size on a column.

Did Gilbert know of Sverdlov's paper? Mirzabekov apparently did not mention it. Gilbert gave a talk on his chemical DNA sequencing method at a meeting in Kiev in September 1975. Sverdlov told me that he was present and that he there gave Gilbert a reprint of his 1973 paper. Maxam and Gilbert did not cite it in their joint paper (1), but Maxam, who eventually became a graduate student, did cite it in his thesis (4). He also cited it in a review he published in 1983 which is essentially a reprint of his thesis (5). There he states that he did all the work with C and T and that it was his idea to use hydrazine. Before he begins this section he writes "It should be noted in passing that Sverdlov *et al.* (1973) had proposed limited base-specific chemical cleavage for sequencing end labelled DNA in 1973".

Was Sverdlov's paper (3) cited by western molecular biologists? The Citation Index allows one to check. The answer is revealing: it was cited just once (once!) by a western author in a journal. The article (6) which quotes it is to the point. It begins: "Limited base-specific or base selected cleavage of a defined DNA fragment yields polynucleotide products, the length of which correlates with the position of the particular base (or bases) in the original fragment. Sverdlov and coworkers recognized the possibility of using this principle for the determination of DNA sequences". So what, one may say! Everybody has ideas. Ideas are nothing. What counts is the realisation of ideas. So one may ask: why did Sverdlov not continue to work on the method when he had the proper idea? The answer is simple. His chief, Ovchinnicov, thought the scheme would never work. He did not allow Sverdlov to continue his work. When I saw Sverdlov for the first time in the late seventies in Moscow, he had a poster hanging in his lab saying *VANITAS VANITATUM*.

It is of some interest to compare the lives of the two protagonists, the winner, Walter Gilbert, and the loser, Eugene Sverdlov. Both had Jewish parents. Sverdlov is about six years younger than Gilbert. Walter Gilbert's father was a professor of economics at Harvard University, Sverdlov's father a communist administrator somewhere in the Ukraine. Sverdlov's mother was a teacher like Gilbert's. All of Sverdlov's relatives perished either in the Gulag, as Soviet soldiers in WW II, or as victims of the *Einsatzkommando D*. In the summer of 1942 these German murderers and their Russian helpers in police uniforms came to the small town near Stavropol in the Northern Caucasus to which Sverdlov's mother had fled. She took the four year old Eugene by the hand and jumped in the pit full of dead Jews before the murderers had time to shoot. Mother and son crawled out of the pit at night and found someone who had the courage to hide them until the Red Army liberated the town. Then they were interrogated again and again. How could they have survived without being traitors? Young Sverdlov wanted to study physics but the Jewish quota was full and was he not possibly a traitor? So he studied chemistry instead. Gilbert had a peaceful youth in Cambridge, Mass. He studied theoretical physics at Harvard. He had complete freedom in James

Watson's lab at Harvard. Sverdlov worked under Ovchinnicov, the great tyrant and charming actor. After Ovchinnicov's death and the breakdown of the USSR, he became the director of an institute in Moscow.

So much about the history of chemical DNA sequence analysis. It remains only to be said that the method was used immediately by a student of Walter Gilbert to sequence the *lac I* (repressor) gene (7). It turned out that 13 of the 360 residues were either missing or had been wrongly determined in the protein sequence. A student of mine then sequenced the *lacY* (permease) gene (8). The techniques of DNA cloning and sequencing were so seductive that another student of mine then cloned and sequenced the *galR* (repressor) gene, without any expectation other than finding another repressor sequence (9). It was a most pleasant surprise when it turned out that the sequence of Gal repressor was homologous to the sequence of Lac repressor! Gal repressor was the second member of a large family of sequence related repressors. Sequencing the *lacZ* gene then just required the effort of an undergraduate and a technician (10). The protein sequence turned out to be almost correct. There were only two mistakes in its 1021 residues. This marked the end of sequencing large proteins.

References

(1) Maxam, A.M. & Gilbert, W.: A new method for sequencing DNA. Proc.Natl.Acad.Sci. USA **74**, 560-564, 1977.

(2) Gilbert, W., Maxam, A.M. & Mirzabekov, A.: Contacts between the *lac* repressor and DNA revealed by methylation. In: Control of ribosome synthesis. Alfred Benson Symposium IX. Ed. by N.C. Kjelgaard and O. Maaloe. Munksgard, Copenhagen, 139-143, 1976.

(3) Sverdlov, E.D., Monastyrskaya, W., Chestukhin, A.V. & Budowsky, E.I.: The primary structure of oligonucleotides. Partial apurination as a method to determine the position of purine and pyrimidine residues. FEBS Letters **33**, 15-17, 1973.

(4) Maxam, A.M.: Thesis: Nucleotide sequence of DNA. Harvard University, 1980.

(5) Maxam, A.M.: Nucleotide sequence of DNA. In: Methods of DNA and RNA sequencing. Ed. by A.M. Weissman. Praeger, New York, 113-164, 1983.

(6) Ambrose, B.J.B. & Pless, R.C.: DNA sequencing: chemical methods. Methods in Enzymology **152**, 522-538, 1987.

(7) Farabough, P.: Sequence of *lac I* gene. Nature **274**, 765-769, 1978.

(8) Büchel, D.E., Gronenborn, B. & Müller-Hill, B.: Sequence of the lactose permease gene. Nature **283**, 542-545, 1980.

(9) Wilcken-Bergmann, B.v. & Müller-Hill, B.: Sequence of *gal R* gene indicates a common evolutionary origin of *lac* and *gal* repressor in *Escherichia coli*. Proc.Natl.Acad.Sci. USA **79**, 2427-2431, 1982.

(10) Kalnins, A., Otto, K., Rüther, U. & Müller-Hill, B.: Sequence of the *lac Z* gene of *Escherichia coli*. EMBO J. **2**, 593-597, 1983.

(11) Steitz, T.A.: Structural studies of protein-nucleic acid interactions. The source of sequence-specific binding. Cambridge University Press, Cambridge, 1993.

1.18 Steric Hindrance is the Mechanism of Repression

DNA methylation can be used for the analysis of specific protein-DNA contacts (1). This was mentioned in the last chapter. Suppose the *lac* operator is occupied by Lac repressor. If Lac repressor recognises the surface of the major groove, the N7's of the guanines (G's) positioned at this interface should not be methylated under conditions where they are methylated in the absence of Lac repressor. Walter Gilbert and his coworkers (1) did this experiment with 01, the main operator of the *lac* system. They found indeed that some G's are protected (Fig. 7). If a particular G is totally protected, the band corresponding to this G will be absent. In a similar manner, the possible protection of A's in the minor groove can be tested. It should be recalled that all four possible base pairs look different in the major groove (see Fig. 6). Thus, interactions in the major groove are of particular interest for specific DNA-protein recognition. In contrast, A-T and T-A look identical in the minor groove. The corresponding is true for G-C and C-G pairs. Thus, in the minor groove the only distinction is between A-T/T-A and G-C/C-G base pairs (see Fig. 6).

The methylation protection technique was extensively used by Ronald Ogata, a postdoctoral fellow in Gilbert's laboratory (2). He demonstrated that a tryptic N-terminal fragment (59 residues long) of Lac repressor was solely responsible for specific protein-DNA recognition of *lac* operator (see section 1.13.).

Gilbert also recognized the potential of ethylnitrosourea. This compound specifically ethylates the phosphates of DNA (3). The rate of hydrolysis of the ethylated phosphate differs from the hydrolysis of unsubstituted DNA. This reaction was used successfully by Gilbert's technician Maxam to mark those phosphates of *lac* operator which are protected, *i.e.* bind to side chains of Lac repressor (4). The same was done by three students of Walter Gilbert with RNA polymerase-*lac* promoter complex (5), the CAP protein-*lac* CAP site complex, and the λ *CI* repressor-operator complex (6). The details of some of these experiments have never been published in journals. There was little time left for writing in this vortex of production.

The DNase footprint experiment was invented in the laboratory of Jeffrey Miller then in Geneva. Its logic is similar to that of the Maxam-Gilbert method. Take DNA which is labelled at one end. Treat it with DNase in the presence and absence of the DNA-binding protein in such a manner that each DNA molecule is cut in the average once. Separate the pieces on an acrylamide gel. The base pairs of a region where a protein is bound specifically will not be cut. One will see a

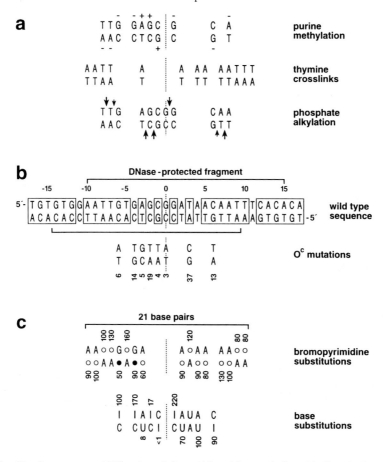

Fig. 7.: The *lac* operator, 1978, adapted from (11), with permission. (a) Chemical modifica-
tions of *lac* operator DNA. The upper portion shows the position at which Lac repressor inhibits (-) or enhances (+) methylation of purines by dimethylsulfate (2). The
middle portion shows the positions of thymines at which repressor crosslinks *BrU*-
substituted DNA upon UV radiation. The lower portion shows the positions at which
repressor inhibits alkylation of phosphates by ethylnitrosourea (3,4). (b) Sequence of
lac operator. DNA fragment protected against DNase action by repressor (8). Base pair
substitution of O^c mutations. The T/G bp in position 4 is in the original! The numbers
below indicate the corresponding affinities for O^c operator expressed as percent of the
affinity for O^+ operator for the tight binding *I′X86* repressor protein (11). (c) Synthetic *lac* operators, O bromouridine, ● bromocytidine. The numbers indicate the corresponding affinities of Lac repressor for modified *lac* operator, expressed as percent of
the affinity for unmodified synthetic *lac* operator, for *I′X86* repressor protein.

blank *window* where no cuts have occurred. The window will be somewhat larger than the DNA binding protein itself, since DNase is somewhat bulky. David Galas and Albert Schmitz did this experiment first with *lac* operator and a tight binding mutant of Lac repressor (8). This is one of the most successful methods to test protein-DNA interactions. One may think it is simply a direct spin off of the Maxam-Gilbert method. This is not so. Albert Schmitz, one of my former graduate students, told me the story of the history of this experiment. Albert liked playing practical jokes. So, one day he put a very small amount of DNase into a sample of radioactively labelled *lac* operator and Lac repressor, which his colleague David Galas had just put in the refrigerator. The next day, Galas ran a gel of this sample. And to his utter surprise he saw many bands with an empty window in the middle. Albert Schmitz confessed. And then they both realized what a marvelous gift from heaven this was: when written down it became one of the most cited papers in molecular biology. It should be added that they quickly tested the general application of the method by testing the binding of RNA polymerase and of the CAP protein in their *lac* binding sites (9,10).

Jacob and Monod argued intuitively that repressor competes with RNA polymerase. Gilbert and his students found "that Lac repressor covers a sequence transcribed into the initial portion of the *lac* messenger. Thus the repressor functions by blocking access of the RNA polymerase to its initiation site" (12). The DNase protection experiments (9) confirmed this work: *lac* promoter and *lac* operator overlap. So the mechanism of repression was simple and easy to understand.

References

(1) Gilbert, W., Maxam, A.M. & Mirzabekov, A.: Contacts between the *lac* repressor and DNA revealed by methylation. In: Control of ribosome synthesis. Alfred Benzon Symposium IX. Ed. by N.C. Kjeldgaard & O. Maaloe. Munksgard, Copenhagen, 139-143, 1976.

(2) Ogata, R.T. & Gilbert, W.: DNA-binding site of *lac* repressor probed by dimethylsulfate methylation of *lac* operator. J.Mol.Biol. **132**, 709-728, 1979.

(3) Sun, L. & Singer, B.: The specificity of different classes of ethylating agents toward various sites fo HeLa cell DNA in vitro and in vivo. Biochemistry **14**, 1795-1802, 1975.

(4) Gilbert, W. & Maxam, A.M.: personal communication. Cited in: Barkley, M.D. & Bourgeois, S.: Repressor recognition of operator and effectors. In: The operon. Ed. by

J.H. Miller & W.S. Reznikoff. Cold Spring Harbor Laboratory, Cold Spring Harbor, NY, 177-220, 1978.

(5) Johnsrud, L.: Contacts between *Escherichia coli* RNA polymerase and a *lac* operon promoter. Proc.Natl.Acad.Sci. USA **75**, 5314-5318, 1978.

(6) Majors, J.E.: Thesis, Harvard University, 1977.

(7) Johnson, A.: Thesis, Harvard University, 1980. Cited in: Jordan, S.R. & Pabo, C.O.: Structure of the λ complex at 2.5 Å resolution: details of the repressor-operator interaction. Science **242**, 893-899, 1988.

(8) Galas, D.J. & Schmitz, A.: DNase footprinting: a simple method for the detection of protein-DNA binding specificity. Nucl.Acids Res. **5**, 3157-3170, 1978.

(9) Schmitz, A. & Galas, D.J.: The interaction of RNA polymerase and Lac repressor with the *lac* control region. Nucl.Acids Res. **6**, 111-137, 1979.

(10) Schmitz, A.: Cyclic AMP receptor protein interacts with lactose operator DNA. Nucl.Acids Res. **9**, 277-292, 1981.

(11) Barkley, M.D. & Bourgeois, S.: Repressor recognition of operator and effectors. In: The operon. Ed. by J.H. Miller & W.S. Reznikoff. Cold Spring Harbor Laboratory, Cold Spring Harbor, NY, 177-220, 1978.

(12) Gilbert, W., Gralla, J., Majors, J. & Maxam, A.: Lactose operator sequences and the action of Lac repressor. In: Protein-Ligand Interactions. H. Sund & G. Blauer eds. de Gruyter, Berlin-New York, 193-206, 1975.

1.19 Chemical Synthesis of *lac* Operator DNA

When the news spread that DNA could be cloned, it became a challenge to chemically synthesize DNA and to clone it. At that time the synthesis of pentamers seemed just possible. One could ligate such pentamers together. But, how could one test for the presence of such an oligomer cloned in plasmid DNA at a time when DNA sequencing was still unsolved? Then it came in handy to recall an old experiment, escape synthesis (1). This experiment demonstrates that excess copies of *lac* operator on heat-induced phage λ, which do not lyse the host cells titrate the ten copies of Lac repressor in a wildtype $I^+Z^+Y^+$ cell. The presence of excess *lac* operator thus relieves repression and leads to constitutive β-galactosidase synthesis. In 1976 two groups (2,3) showed that this test could be applied to demonstrate the presence of synthetic *lac* operator DNA, which was 22 base pairs long and cloned on the multicopy plasmid pBR 322.

Synthetic *lac* operators of various lengths could also be tested for their functioning *in vitro* with the nitrocellulose filter binding assay. Lac repressor like most proteins binds tightly to nitrocellulose. If one labels *lac* operator with radioactive ^{32}P, one can measure the amount of label which sticks to the nitrocellulose filter (4). This was done with *lac* operators which were 16, 17, 19 or 26 base pairs long. They bound with 1.5%, 12%, 40% or 100% efficiency respectively (5).

The chemistry of DNA synthesis was further refined by Marvin Caruthers. He and his collaborators synthesized *lac* operators containing base analogues such as uracil (thymine without a methyl group!), hypoxanthine (guanine missing the NH_2 group in position 2 pointing to the minor groove!) or bromouracil (which can react with protein after UV radiation). The conclusion from these studies was straightforward: Lac repressor binds to two major grooves on one side of the DNA (6,7).

References

(1) Revel, H.R. & Luria, S.E.: On the mechanism of unrepressed galactosidase synthesis controlled by transducing phage. Cold Spring Harbor. Symposia of Quantitative Biology **28**, 403-407, 1963.

(2) Heyneker, H.L., Shine, J., Goodman, H.M., Boyer, H.W., Rosenberg, J., Dickerson, R.E., Narang, S.A., Itakura, K., Lin, S.Y. & Riggs, A.D.: Synthetic *lac* operator DNA is functional *in vivo*. Nature **263**, 748-752, 1976.

(3) Marians, K.J., Wu, R., Stawinski, J., Hozumi, T. & Narang, S.A.: Cloned synthetic *lac* operator DNA is biologically active. Nature **263**, 744-748, 1976.

(4) Bahl, C.P., Wu, R., Itakura, K., Katagiri, N. & Narang, S.A.: Chemical and enzymatic synthesis of lactose operator of *Escherichia coli* and its binding to lactose repressor. Proc.Natl.Acad.Sci. USA **73**, 91-94, 1976.

(5) Bahl, C.P., Wu, R., Stawinsky, J. & Narang, S.A.: Minimal length of the lactose operator sequence for the specific recognition by the lactose repressor. Proc.Natl.Acad.Sci. USA **74**, 966-970, 1977.

(6) Goedel, D.V., Yansura, D.G. & Caruthers, M.H.: Binding of synthetic lactose operator DNAs to lactose repressors. Proc.Natl.Acad.Sci. USA **74**, 3292-3296, 1977.

(7) Goedel, D.V., Yansura, D.G. & Caruthers, M.H.: How *lac* repressor recognizes *lac* operator. Proc.Natl.Acad.Sci. USA **75**, 3578-3582, 1978.

1.20 Lac Repressor-Operator Complex Binds to Nitrocellulose

Mark Ptashne and Walter Gilbert had used glycerol gradients to show the specific interaction between Lac and λ *CI* repressor and their operators, which were embedded in λ DNA (see section 1.11.). The method worked well but it was laborious and time-consuming. It took a day and needed an ultracentrifuge to test six samples. It was, therefore, most welcome when Arthur Riggs and Suzanne Bourgeois (a former student of Jacques Monod, who had followed her husband Melvin Cohn to the States) developed a new method which was at least an order of magnitude faster: the nitrocellulose filter binding assay. Lac repressor binds like most but not all proteins to nitrocellulose, whereas double stranded DNA does not bind. If one filters a mixture of radioactively labelled λ DNA containing *lac* operator and Lac repressor through a nitrocellulose filter, the repressor-operator complex will stick to the filter, but free DNA will pass through. The labelling of the λ *lac* DNA had to be done by growing the λ lysogenic bacteria in the presence of radioactive phosphate.

Riggs, Bourgeois and Cohn, set out to determine first the equilibrium of the repressor-operator complex (1). At the low salt concentrations they used, it was about 10^{-13} M. They then examined whether Lac repressor had the same specificity towards inducers *in vitro* as *in vivo*. The specificity was the same (2-4). They then proceeded to study the kinetics of repressor-operator interaction. The forward rate constant was astonishingly fast, 7×10^9 M^{-1} sec^{-1}, and the backward rate constant was reasonably slow: the half life of the *lac* repressor-operator complex was about 20 minutes (5). Did the repressor actually use one dimensional diffusion to find its target as Max Delbrück had proposed (6)? This question could be raised but not answered. Superficially, all results seemed to fit the expectations derived from the *in vivo* work. O^c operator bound less well than O^+ operator (7). There were some slight inconsistencies, but the general picture seemed perfect (8). Later we will see (section 2.13.) that the field is full of pitfalls.

References

(1) Riggs, A.D., Suzuki, H. & Bourgeois, S.: Lac repressor-operator interaction I. Equilibrium studies. J.Mol.Biol. **48**, 67-83, 1970.

(2) Riggs, A.D., Newby, R.F. & Bourgeois, S.: Lac repressor-operator interaction II. Effect of galactosides and other ligands. J.Mol.Biol. **51**, 303-314, 1970.

(3) Jobe, A. & Bourgeois, S.: Lac repressor-operator interaction VI. The natural inducer of the *lac* operon. J.Mol.Biol. **69**, 397-408, 1972.

(4) Jobe, A. & Bourgeois, S.: Lac repressor-operator interaction VIII. Lactose is an anti-inducer of the *lac* operon. J.Mol.Biol. **75**, 303-313, 1972.

(5) Riggs, A.D., Bourgeois, S. & Cohn, M.: The *lac* repressor-operator interaction III. Kinetic studies. J.Mol.Biol. **53**, 401-417, 1970.

(6) Adam, G. & Delbrück, M.: Reduction of dimensionality in biological diffusion processes. In: Structural chemistry and molecular biology. Ed. by A. Rich & N. Davidson. Freeman, San Francisco, 198-215, 1968.

(7) Jobe, A., Sadler, J.R. & Bourgeois, S.: Lac repressor-operator interaction IX. The binding of *lac* repressor to operators containing O^c mutations. J.Mol.Biol. **85**, 231-248, 1974.

(8) Barkley, M.D. & Bourgeois, S.: Repressor recognition of operators and effectors. In: The operon. Cold Spring Harbor Monograph Series. Cold Spring Harbor, N.Y., 177-220, 1978.

1.21 Polyacrylamide Gel Electrophoresis of Protein-DNA Complexes

DNA restriction fragments carrying *lac* operator (01) or the CAP site form under suitable conditions specific complexes with Lac repressor or CAP protein. The DNA concentration has to be low, about 10^{-9} to 10^{-10}M. The salt concentration of the buffer also has to be low. Free DNA and DNA-repressor or CAP complex can be separated on polyacrylamide gels. Mutants in the operator or the CAP target can be used as specific controls. At first glance it is astonishing that the DNA-protein complexes are stable enough in the gel to be separated from the free DNA. The protein-DNA complexes behave as if they were stabilized in a small cage. This most useful technique was discovered by two groups: Mark Garner and Arnold Revzin at Michigan State University used CAP protein and a 203 bp restriction fragment carrying the *lac* promoter region with its CAP site (1). They could show that the CAP-complex forms only in the presence of cAMP. It does not form with a DNA carrying the L8 mutation in the CAP site, which *in vivo* drastically reduces CAP activation. Here, the experiments were straightforward in their interpretation.

The same cannot be said for the experiments of the other group, Michael Fried and Donald Crothers of Yale University (2). They used the same 203 base pair restriction fragment Garner and Revzin used. But they analysed binding of Lac repressor to *lac* operator 01. With increasing concentrations of Lac repressor additional bands appeared on the gel which moved slower and slower. Altogether, they observed eight (!) bands. In retrospect, it seems that the DNA concentration they used was much too high. Apparently they analysed nonspecific binding. They then cut the 203 bp fragment into two pieces, one carrying 01 and the other 03, and tried to measure repressor binding to each. Their analysis of the data is misleading. They concluded that 03 is 18-fold weaker than 01. In reality, it is 300-fold weaker. They further concluded that 01 and 03 do not influence each others binding to Lac repressor. In reality, Lac repressor forms tight loops with 01 and 03 (see section 3.4.2.). Irrespective of the fact that for years their particular conclusions were misleading (see for example 3,4), the invention of gel electrophoresis as a test for protein-DNA complexes is truly one of the great discoveries in the field. Today, it helps everybody who works with DNA binding proteins from eukaryotes tremendously.

References

(1) Garner, M.M. & Revzin, A.: A gel electrophoresis method for quantifying the binding of proteins to specific DNA regions: application to components of the *Escherichia coli* lactose operon system. Nucl.Acids Res. **9**, 3047-3060, 1981.

(2) Fried, M. & Crothers, D.M.: Equilibria and kinetics of *lac* repressor-operator interactions by polyacrylamide gel electrophoresis. Nucl.Acids Res. **9**, 6506-6525, 1981.

(3) Crothers, D.M.: Gel electrophoresis of protein-DNA complexes. Nature **325**, 464-465, 1987.

(4) Gerstle, J.T., Fried, M.: Measurement of binding kinetics using the gel electrophoresis mobility shift assay. Electrophoresis **14**, 725-731, 1993.

1.22 Fusion of and Complementation within the *lacZ* Gene

Three percent of the soluble protein of *E. coli* is present as β-galactosidase in *lac* constitutive mutants. This invited extensive biochemical studies of this enzyme (1,2). It is a tetramer consisting of four identical subunits. Its proteolytic breakdown obscured its large subunit molecular weight for some years. Each subunit has a molecular weight of about 120,000 Daltons (3,4). In *E. coli* its MW is surpassed only by the MWs of the β and β' subunits of RNA polymerase. The determination of the primary structure of β-galactosidase thus became the Mount Everest of protein chemistry. It was conquered by Audrey Fowler and Irving Zabin just at the time when DNA sequencing became known (5). It is thus one of the rare cases in which all of the protein sequence and only a small part of the DNA sequence were known when it was published. When the DNA sequence was solved (6), it was remarkable and most pleasant to see that almost all of the protein sequence was correct.

β-galactosidase is active only as a tetramer. Agnes Ullmann had shown that mutations in some of the codons coding for the N-terminal 60 or C-terminal 100 residues result in inactive, dimeric β-galactosidase (7). In the case of such internal deletions, the missing peptide can be replaced by the corresponding N-terminal or C-terminal peptides. This can be done *in vivo* or *in vitro*. The added peptides fuse together two inactive dimers to become active tetramers. This is called intracistronic alpha or omega complementation. Thus, in the proper background, the presence of a 60 residue long alpha peptide results in easily measurable β-galactosidase activity. Later we will see that this property has been used extensively in gene cloning.

Another property of β-galactosidase is the fact that its N-terminal 23 amino acid residues can be replaced by any amino acid residues without influencing the enzymatic activity of β-galactosidase. This was first demonstrated through the existence of fusions between active Lac repressor and active β-galactosidase (8). They were obtained as *lac*⁺ revertants of *IᵠZ⁻ᵁ¹¹⁸* bacteria. The *U118* mutation is an ochre mutation in the codon of the Z gene coding for residue 17 of β-galactosidase. The revertants may fuse short pieces of Lac repressor, or also the entire functioning Lac repressor to β-galactosidase. These repressor-fusions are interesting in themselves. If one assumes that tetrameric β-galactosidase is assembled in the usual way then one has to assume that the repressor tetramer is separated into two sets of functioning dimers. Proof for this was presented by experiments with cross-linking chemicals (10). The capacity to form N-terminal fusions was later extensively used and exploited by Jon Beckwith and his students (11). If one can place the gene of interest upstream of β-galactosidase, one can isolate any fusion-protein with β-galactosidase as reporter enzyme. Of course, this became even easier with the advent of *in vitro* techniques. This will be dealt with later.

References

(1) Wallenfels, K. & Malhotra, O.P.: Galactosidases. Advances in Carbohydrate Chemistry **16**, 239-298, 1961.

(2) Zabin, I. & Fowler, A.V.: β-galactosidase and thiogalactosidase transacetylase. In: The lactose operon. Ed. by J.R. Beckwith & D. Zipser. Cold Spring Harbor Laboratory. Cold Spring Harbor, N.Y., 27-48, 1970.

(3) Cohn, M.: Contributions of studies of β-galactosidase of *Escherichia coli* to our understanding of enzyme synthesis. Bact.Rev. **21**, 140-168, 1957.

(4) Wallenfels, K., Sund, H. & Weber, K.: Die Untereinheiten der β-Galaktosidase aus *E. coli*. Biochem.Z. **338**, 714-727, 1963.

(5) Fowler, A. & Zabin, I.: Amino acid sequence of β-galactosidase. J.Biol.Chem. **253**, 5521-5525, 1978.

(6) Kalnins, A., Otto, K., Rüther, U. & Müller-Hill, B.: Sequence of the *lac Z* gene of *E. coli*. EMBO J. **2**, 593-597, 1983.

(7) Ullman, A. & Perrin, D.: Complementation in β-galactosidase. In: The lactose operon. Ed. by J.R. Beckwith & D. Zipser. Cold Spring Harbor Laboratory, Cold Spring Harbor, N.Y., 143-172, 1970.

(8) Müller-Hill, B. & Kania, J.: Lac repressor can be fused to β-galactosidase. Nature **249**, 561-563, 1970.

(9) Zabin, I., Fowler, A.V. & Beckwith, J.R.: Position of the β-galactosidase ochre mutant U118. J.Bact. **133**, 437-438, 1978.

(10) Kania, J. & Brown, D.T.: The functional repressor parts of tetrameric lac repressor-β-galactosidase chimera are organised as dimers. Proc.Natl.Acad.Sci. USA **73**, 3529-3533, 1976.

(11) Bassford, P., Beckwith, J., Berman, M., Brickman, E., Casadaban, M., Guarente, L., Saint-Girons, I., Sarthy, A., Schwartz, M., Shuman, H. & Silhavy, T.: Genetic fusions of the *lac* operon: a new approach to the study of biological processes. In: The operon. Ed. by J.H. Miller & W.S. Reznikoff. Cold Spring Harbor Laboratory, Cold Spring Harbor, N.Y., 245-262, 1978.

1.23 Isolation of Lac Permease

The discovery of a transport system which accumulates galactosides inside *E. coli* and which is coded for by the *Y* gene, which is part of the *lac* operon (1,2), was a major discovery. Henceforth, a transport process could be analysed biochemically and genetically. The isolation of the product of the *lac Y* gene became a major project. It took ten years before Fred Fox and Eugene Kennedy solved the problem (3). They had observed that Lac permease can be irreversibly inhibited by N-ethyl-maleimide which presumably alkylates one or several permease cysteine SH groups. The presence of thio-digalactoside, a substrate of Lac permease, protects against inhibition and alkylation. Thus, Fox and Kennedy treated *lacY* constitutive *E. coli* cells first with thio-digalactoside and then with N-ethyl-maleimide. Now all other reactive SH-groups of *E. coli* were alkylated. Then they freed the *E. coli* cells first from the N-ethyl-maleimide and then from the thio-digalactoside. When they now treated these cells with radioactively labelled N-ethyl-maleimide, only Lac permease is alkylated. Inactive, specifically labeled Lac permease was purified in the presence of SDS or similar detergents. This permease is a highly lipophilic protein of the apparent subunit molecular weight of about 30,000 Daltons. It was without doubt the product of the *lacY* gene, Lac permease, which Fox and Kennedy had isolated. To document that it was their brain-child, they gave it a new name: M (membrane) protein. When you leave this world, the objects you named still exist and may be talked about. It was a long way to go from here to solve its structure and mechanism of function, but it provided the confidence that this could be done (4).

References

(1) Cohen, G.N. & Rickenberg, H.V.: Étude directe de la fixation d'un inducteur de la β-galactosidase par les céllules d'*Escherichia coli*. Compt.Rendus **240**, 466-468, 1955.

(2) Rickenberg, H.V., Cohen, G.N., Buttin, G. & Monod, J.: La galactoside perméase d'*Escherichia coli*. Ann.Inst.Pasteur **91**, 829-857, 1956.

(3) Fox, C.F. & Kennedy, E.P.: Specific labelling and partial purification of the M protein, a component of the β-galactoside transport system of *Escherichia coli*. Proc.Natl.Acad.Sci. USA **54**, 891-899, 1965.

(4) Kennedy, E.P.: The lactose permease system of *Escherichia coli*. In: The lactose operon. Ed. by J.R. Beckwith & D. Zipser. Cold Spring Harbor Laboratory Press, Cold Spring Harbor, N.Y. 49-92, 1970.

1.24 The End of an Era

All things come to an end and in science ends come fast. Bacterial control systems were for three decades the only control systems one could analyse efficiently. The *lac* system became the paradigm among very few others. This was made possible by bacterial genetics. Mutations could be ordered on the bacterial chromosome within a very few base pairs, and analysed for their dominance. The methods of genetics were as highly developed as possible. Bacterial genetics combined with modern biochemistry seemed unbeatable.

But in the seventies radically new methods and techniques were developed. Restriction enzymes were discovered. They allowed the ordering and cloning of pieces of DNA. DNA could be sequenced. DNA could be synthesized. After the relevant genes were sequenced, it seemed as if all the problems of gene control in bacteria had been solved. Only a very few geneticists stayed with *lac* and bacteria. The future for bright young people was elsewhere, in eukaryotes.

Indeed, the new methods and techniques produced more than bacterial genetics could ever produce, and transcription control in eukaryotes is much more complicated and thus much more interesting. For the bacterial geneticists it was painful to desert the old techniques when the field became deserted. New developments with prokaryotes were no longer included in textbooks. Money and interest had moved elsewhere. When in 1969 the conference on the *lac* operon was held at Cold Spring Harbor no sign of the new era could be seen. When, seven years later, the conference on the operon was held at the same place, the new era was already in full swing. It is symbolic that Jacques Monod died in 1976, some months before this conference, which he had planned to attend. And it is equally symbolic that Walter Gilbert did not have the time to attend this conference or write a paper for it. His new technique of DNA sequencing was used and quoted by others before he even published it (1)! He was elsewhere and the action has been elsewhere ever since.

Monod was the driving force in the elucidation of the *lac* system. It is thus appropriate to end this chapter by remembering him. In 1971, he published a book in which he tried to tell nonscientists the essence of science (2). What could the general public learn from the analysis of the *lac* system? The title of his book is beautiful: "Chance and Necessity". On page one he writes that the title is taken from a sentence of Democritus: "All that exists in the universe is the fruit of chance and necessity." This sounds true but neither Democritus nor any other Greek philosopher ever said or wrote either this, or anything even vaguely similar. By chance

I found a possible source. Goethe wrote in his *"Wilhelm Meisters Lehrjahre"*: "The matter of this world is formed by necessity and chance" (3). It is worthwhile to continue reading Goethe for he had more to say on this subject, but this chapter is about Monod, not about Goethe. In the last chapter of his book Monod draws his conclusions from his study of the *lac* system. He pleads for an ethic of knowledge. He names as the highest human qualities those which were so often his own: courage, altruism, generosity and creative drive. These should be the highest values for future generations of scientists, readers will agree. And what about the dark sides (4)? And where do these values come from? Here Monod was ambiguous. Some of his last words were: "Je cherche à comprendre", I try to understand (5) is a simple translation. The moment you look into a dictionary of synonyms you see an universe.

References

(1) For example Fig. 5 of Barkley, M.D. and Bourgeois, S.: Repressor recognition of operator and effectors. In: The operon. Ed. by J.H. Miller and W.S. Reznikoff. Cold Spring Harbor Laboratory, 117-220, 1978.

(2) Monod, J.: Le hasard et la nécessité. Éditions du Seuil, Paris, 1970.

(3) Goethe, W.v.: Wilhelm Meisters Lehrjahre. Volume 1, chapter 17. First edition, 1795.

(4) Stanier, R.Y.: The outer and the inner man. In: Origins of molecular biology. A tribute to Jacques Monod. Ed. by A. Lwoff & A. Ullmann. Academic Press, New York, San Francisco, London. 25-29, 1978.

(5) Brunerie, M.: Once upon a time. In: Origins of molecular biology. A tribute to Jacques Monod. Ed. by A. Lwoff & A. Ullmann. Academic Press, New York, San Francisco, London. 37-42, 1979. Agnes Ullmann (letter of Nov.21,1994 to BMH) reports that Monod said this sentence to his doctor, Gerard Mantoux.

Part 2

Misinterpretation

2.1 The Interpretation of Experiments is Sometimes Wrong

Now is the proper time to reflect. Today it is pretty clear what was right and what was occasionally wrong with these old experiments dealing with the *lac* system. In the years gone by, the right solutions were used over and over again for new predictions and new experiments. The erroneous interpretations of some experiments were simply forgotten. They could not be used for further experiments. The history of the *lac* system thus provides an excellent example for following the progress of the science of genetics at a particular object. I note that today two opposing views are held on the progress of science:

1. Probably all productive scientists think that the scientific endeavor brings them closer to truth. This view is today shared only by a minority of philosophers and sociologists of science. The best known of these is the late Karl Popper (1).
2. These days more and more sociologists and historians of science write that in science as in politics, subjective constructions are proposed, judged and decided according to social pressure. I cite here the names of Paul Feyerabend (2), Bruno Latour (3) and Karin Knorr-Cetina (4) as some of the proponents of this view.

The history of the *lac* system can be used to disprove the latter proposition. It begins in the thirties and forties with the observations of Henning Karström and Jacques Monod on the "adaptation" of *E. coli* to sugars such as lactose. At the beginning of a new branch of science there is often total confusion. Too many possibilities can be used to explain the initial observations. It may take years until the proper experimental techniques are available to decide which of the explanations is correct. In the early fifties the chemical synthesis of certain organic chemicals (inducers and substrates) opened the way to decide unambiguously whether *de novo* synthesis or "instruction" (activation) of pre-enzymes is the proper explanation for the phenomenon. In the late fifties, mutant repressors and operators were used to demonstrate negative control, and to exclude all other possible models. I will show that at some of these crossroads, some scientists went off in the wrong direction and with scanty evidence backed a wrong explanation. How could this happen? The interpretation of experiments often depends upon tacit assumptions. Most of the time these assumptions are reasonable. But sometimes it may take years until one discovers that such an assumption was wrong.

It is easy to add more of the same to old knowledge, but it is difficult to explain the new. I was astonished to learn how many papers dealing with the *lac* system actually are quite wrong, not because the experiments themselves were faulty, but because some of the assumptions used in their interpretation turned out to be wrong. In retrospect it is difficult to understand how things could go so wrong. But at the time those possibilities seemed real which now we consider as totally unlikely and even absurd.

Usually one forgets those cases. They are not mentioned in textbooks or reviews. Yet it seems to me that students would profit from a discussion of specific cases. Therefore, I include here a section on misinterpretations which later were clarified unambiguously. Most of the experiments are old. They were performed during the time period described in the first part of this book. A very few examples are newer. Of course, there will be more. Often it will simply take some time until misinterpretations are generally agreed upon. Science seems to advance as does evolution: in small steps, correcting one mistake after the other, always becoming better and better, but evidently lacking a central grand plan or vision.

References

(1) Popper, K.: Conjectures and refutations. Routledge & Kegan Paul, London, 1963.
(2) Feyerabend, P.: Against method. New Left Books, London, 1975.
(3) Latour, B. & Woolgar, S.: Laboratory life. The social constructions of scientific facts. Sage, Beverly Hills, 1979.
(4) Knorr-Cetina, K.: The manufacture of knowledge. An essay on the constructivist and contextual nature of science. Pergamon Press, Oxford, 1981.

2.2 Adaptation Explained by Self-Replicating Genes

How can the phenomenon of adaptation, or enzyme induction, as Monod preferred to call it, be explained? Sol Spiegelman, who worked on the induction of the galactose metabolizing enzymes in yeast (1) proposed in 1947 the "plasmagene" hypothesis for yeast (2), in which the substrate would turn on enzyme synthesis by inducing multiple duplications of the genes coding for the relevant enzymes. More enzymes would be made because more genes were made. At the time it was generally assumed that genes were also proteins. Perhaps genes and enzymes which were induced were identical. The hypothesis predicted that these genes would occur in the cytoplasm and not in the nucleus, and furthermore it predicted rather special kinetics of enzyme induction. In the case of the *lac* system the hypothesis was soon ruled out (3).

It is worthwhile adding a few sentences about Sol Spiegelman (1914-1983). As a student he went to Spain and fought against Franco. Later, he became a student of Carl Lindegren (4), one of the founders of Neurospora and yeast genetics and a creative iconoclast. I mention here only his book on Lamarckian inheritance (5). During his life Spiegelman was wrong in several instances. He clashed with Charles Weissmann on the mechanism of replication of the RNA phage $Q\beta$. He claimed to have shown that $Q\beta$ + strands directly produce identical $Q\beta$ + strands. He took his defeats in good humor. "Data no longer available" was once his answer in the discussion after a seminar.

But he *was* the co-discoverer of nucleic acid, *i.e.* DNA-RNA, hybridisation (6), which is the basis of some of the most important techniques in molecular biology (*e.g.* Southern, Northern and PCR). He was in Paris when it was announced that Monod would receive the Nobel Prize. That day he wept in the house of Jekisiel Szulmeister. It was all in vain, he would never get the Prize. Spiegelman's attitude towards science is epitomized in a remark he once made in Stockholm, shortly after his great success, the discovery of nucleic acid hybridisation. It was at a dinner in his honor. All evening he had not said a single word. A student tried to cheer him up by praising him as a creative scientist. Then he exploded. "What nonsense, only charlatans are creative" (7).

References

(1) Spiegelman, S. & Dunn, R.: Interactions between enzyme-forming systems during adaptation. J.Gen.Physiol. **31**, 153-173, 1947.

(2) Spiegelman, S. & Reiner, J.: The formation and stabilisation of an adaptive enzyme in the absence of its substrate. J.Gen.Physiol. **31**, 175-193, 1947.

(3) Monod, J., Pappenheimer, A.M. & Cohen-Bazire, G.: La cinéthique de la biosynthèse de la β-galactosidase chez *E. coli* considerée comme fonction de la croissance. Biochim.Biophys.Acta **9**, 648-660, 1952.

(4) Mortimer, R.K.: Lindegren, C.C., iconoclastic father of *Neurospora* and yeast genetics. In: The early days of yeast genetics. Ed. by M.N. Hall & P. Linder. Cold Spring Harbor Laboratory Press, Cold Spring Harbor, N.Y. 17-38, 1993.

(5) Lindegren, C.C.: Cold War in biology. Planarian, Ann Arbor, 1966.

(6) Hall, B.D. & Spiegelman, S.: Sequence complementarity of T2-DNA and T2-specific RNA. Proc.Natl.Acad.Sci. USA **47**, 137-146, 1961.

(7) Klein, G.: Are scientists creative? In: The atheist and the holy city. Encounters and reflections. The MIT Press, Cambridge, Mass., 37-52, 1990. Spiegelman told another story shortly before his death to Hanna Engelberg-Kulka which may be even more revealing. It was about his first experiment. He was a young boy. He wanted to know whether G. existed. His father was a rabbi. So he lit a fire at Yom Kippur. If G. existed he would die. The experiment worked, he survived, G. did not exist, he said. It was an experiment without controls, said Hanna.

2.3 The Kinetics of Adaptation

The very early work on the *lac* system had demonstrated two extreme forms of adaptation of *E. coli* to its surroundings. In 1907 Rudolf Massini had shown that a certain strain (*E. coli mutabile*) was unable to grow on lactose (1). Plated on lactose indicator plates it yielded only *lac* negative colonies. With time, every colony produced papillae on its surface. These papillae consist of bacteria which are *lac*$^+$. Massini correctly concluded that they were rare mutants. These experiments were repeated and confirmed by several scientists and particularly by Jacques Monod in 1946 (2). Monod had also confirmed and extended (3) earlier work by Henning Karström (4) who had shown that it takes some hours before a culture of *E. coli* adapts to a particular sugar it can degrade, for example lactose.

In 1946 the British physical chemist Sir Cyril Hinshelwood (1897-1967) published a book (5) in which he tried to show that chemical reaction kinetics may illuminate the process of bacterial growth and adaptation. Hinshelwood imagined a bacterial cell as the sum of all its chemical reactions which can be pushed in various directions. For example he proposed the equation (p. 22):

enzyme + substrate = more enzyme + products

This equation enchanted him. He believed it to be real. The mutants of Massini (Hinshelwood calls them variants) were for him just the extreme end of an equilibrium of reactions which has at its other end adaptation as analysed by Karström. He thought the mutants of the geneticists were misinterpretations which he put back in their proper frame.

At the time these ideas were shared by many people. They would not merit mentioning here had not Hinshelwood set out to publish a revised version of his book twenty years later (6). In the meantime a voluminous literature on bacterial genetics accumulated. Monod and Jacob had published their fundamental papers on *lac* and λ regulation (7). Hinshelwood's new book had about twice the volume of the original. Avery, Luria and Delbrück, who were omitted in the first edition were now cited. The Jacob and Monod papers were discussed. But Hinshelwood remained firm. He stuck to his old view that Massini, Monod and all of those who repeated these and similar experiments did not observe mutations in the sense that geneticists describe mutations, but instead bacterial variants at the extreme end of an equilibrium.

For Hinshelwood the repressor did not exist. It was a misinterpretation: "If we accept the view that all other effects on the cell of a nonmetabolizable inducer are entirely trivial and secondary, then a purely negative action of this kind would follow. This would mean, however, that the regulatory gene had no real function other than to cut off a normally quite unnecessary process, namely the formation of an inducible enzyme. If it had no other function, then natural selection would seem to have done its work very badly, leaving two genes with no function but to frustrate one another. Such a state of affairs would be surprisingly wasteful" (6, p. 248); and: "… quite apart from this general consideration, it is doubtful whether the statistical evidence about the occurence of positive and negative strains in the above sense is abundant enough to establish the repressor gene as a positive entity, even if the characterization of positive and negative were less arbitrary than it frequently is" (6, p. 282). Hinshelwood's book appeared in 1966. This was the year Lac repressor was isolated (8). Hinshelwood died one year later.

References

(1) Massini, R.: Über einen in biologischer Beziehung interessanten Kolistamm (Bacterium coli mutabile). Arch.f.Hygiene **61**, 256-299, 1907.

(2) Monod, J. & Audureau, A.: Mutation et adaptation enymatic chez *Escherichia coli – mutabile*. Ann.Inst.Pasteur **72**, 868-879, 1946.

(3) Monod, J.: Recherches sur la croissance des cultures bactériennes (Thèse Doctorat ès Sciences). Hermann Ed., Paris, 1942.

(4) Karström, H.: Enzymatische Adaption bei Mikroorganismen. Ergebnisse der Enzymforschung **7**, 350-376, 1938.

(5) Hinshelwood, C.N.: The chemical kinetics of the bacterial cell. Clarendon Press, Oxford, 1946.

(6) Dean, A.C.R. & Hinshelwood, C.: Growth, function and regulation in bacterial cells. Clarendon Press, Oxford, 1966.

(7) Jacob, F. & Monod, J.: Genetic regulation mechanisms in the synthesis of proteins. J.Mol.Biol. **3**, 318-356, 1961.

(8) Gilbert, W. & Müller-Hill, B.: Isolation of the *lac* repressor. Proc.Natl.Acad.Sci. USA **56**, 1891-1898, 1966.

2.4 Lac Repressor is RNA

The PaJaMo experiment (the cross between *Hfr I⁺Z⁺* and female *LacI⁻Z⁻* bacteria) had provided evidence in favor of *negative* control of the *lac* operon (section 1.5.). It was tempting to exploit this test system further to determine the chemical nature of Lac repressor. Was it protein or RNA? One could add an inhibitor of protein synthesis to the mating bacteria and observe quantitatively the possible inhibition of the repression of the synthesis of β-galactosidase. This was done with 5-methyl-tryptophan (1). If one adds sufficient amounts of 5-methyl-tryptophan as inhibitor at the moment mating begins neither β-galactosidase nor Lac repressor can be made. If one adds L-tryptophan 75 minutes later, one can test for the presence of Lac repressor by measuring β-galactosidase synthesis in the presence and absence of inducer during the next half hour. Then one indeed observes repression almost immediately. Pardee and Prestidge did a similar experiment with chloramphenicol (which they did not document) with similar results. They explained the presence of Lac repressor by arguing that it demonstrated that Lac repressor is not a protein but RNA.

What is wrong with this interpretation? In PaJaMo-type experiments, it takes more than an hour for repression to begin. This does *not* imply that the synthesis of Lac repressor is very slow. It indicates that one molecule of mRNA for repressor is synthesized only once an hour in every cell. It takes almost an hour until one molecule of mRNA is made *in each cell* which can then be translated within minutes into Lac repressor. Thus, Lac repressor mRNA may accumulate during 5-methyltryptophan inhibition. It may be translated the moment L-tryptophan is added. Finally, it should be noted that the authors thank Gunther Stent (2) for suggesting the experiment.

It is noteworthy that seven years later, when nonsense mutations had already been found both in the λ *CI* and the *lacI* gene (section 1.8.), Jeffrey Miller presented a model of a putative DNA-RNA triple helix complex between *lac* operator and Lac repressor RNA (3). This was his first paper. The picture of the model looked so seductively beautiful that no evidence was hard enough to disprove it.

References

(1) Pardee, A.B. & Prestidge, L.S.: On the nature of the repressor of β-galactosidase synthesis in *E. coli*. Biochim.Biophys.Acta **36**, 545-547, 1959.
(2) For Gunther Stent see next chapter!
(3) Miller, J.H. & Sobell, H.M.: A molecular model for gene repression. Proc.Natl.Acad.Sci USA **55**, 1201-1205, 1966.

2.5 Operator is RNA or Protein

Jacob and Monod used the principle of Occam's razor when they proposed that repressor interacts with operator DNA to inhibit the start of transcription of the RNA of the operon. This was the simplest, most economical solution of the problem. One could, of course, forget Occam's razor and propose that not transcription but translation of the *lac* operon is actually inhibited by Lac repressor. This was indeed proposed in 1965 by two Dutch outsiders in a rather esoteric journal (1). The proposal was read by two American biochemists. Circumstantial evidence for this model was presented by them one year later at the Cold Spring Harbor Symposium in 1966 (2). In those days, Walter Gilbert and I had just done the experiments which proved that we had isolated Lac repressor. We drove to Cold Spring Harbor to meet the participants of the symposium. We arrived too late to attend any lectures. The symposium coincided with Francis Crick's fiftieth birthday. So, at the end of the last session, his birthday was celebrated. He had to open a gigantic box: out of it came a naked woman. "Francis likes beautiful models", was Jim Watson's comment. A year later Walter Gilbert proved conclusively that Lac repressor binds to *lac* operator DNA (3). This was the end of the ugly model.

At that time Gunther Stent was *the* philosopher-theoretician of molecular biology. He could use the terms epistemology, phenomenology or structuralism quickly and without hesitation. Together with a French post-doc, François Lacroute, as late as 1968 he published a paper, presenting evidence that functional *lac* operator is indeed identical with the extreme N-terminus of β-galactosidase (4). What was his evidence? Suppose you inhibit the start of translation of β-galactosidase with trimethoprim, an inhibitor of translation, then the extent of repression should be influenced if the operator is a peptide. In fact, during such inhibition repression decreases threefold at low inducer concentrations, whereas no difference in repression is observed with constitutive versus fully repressed synthesis of β-galactosidase. The authors had forgotten that Pardee and Prestidge had come to the opposite conclusion with other inhibitors of protein synthesis (see the preceding chapter!). They did not consider the possibility that the concentration of the very few molecules of Lac repressor may decrease during inhibition of protein synthesis and that this may explain their observations. Thus they concluded: "the results presented here are thus compatible with the idea … of a 'regulatory' polypeptide" (4).

At the same time Gunther Stent had come to the conclusion that the end of progress of the arts *and* science was in sight. To him all major problems seemed solved particularly in molecular biology. Nothing significant could be done anymore. A new "golden age" was around the corner. Everybody young in the US would become a hippie with sufficient money for drugs and sex. Stent published his vision in 1969 (5) and then again in 1978 (6), after the invention of DNA cloning and sequencing, with some additions but without changing his conclusions. How wrong can a philosopher of science be! Old age of molecular biology will come, perhaps in a hundred years from now. I presume this will bring confusion but no happiness.

References

(1) Gruber, M. & Campagne, R.N.: Regulation of protein synthesis: an alternative to the repressor-operator hypothesis. Proc.kon.ned.Akad.Wet.Ser.C. **68**, 270-276, 1965.

(2) Cline, A.L. & Bock, R.M.: Translational control of gene expression. Cold Spring Harbor Symp.Quant.Biol. **31**, 321-333, 1966.

(3) Gilbert, W. & Müller-Hill, B.: The *lac* operator is DNA. Proc.Natl.Acad.Sci. USA **58**, 2415-2421, 1967.

(4) Lacroute, F. & Stent, G.S.: Stimulation of the differential rate of β-galactosidae synthesis in *E. coli* by trimethoprim. Mol.Gen.Genet. **101**, 368-375, 1968.

(5) Stent, G.S.: The coming of the golden age: a view of the end of progress. The Natural History Press, Garden City, NY, 1969.

(6) Stent, G.S.: Paradoxes of progress. W.H. Freeman and Company, San Francisco, 1978.

2.6 Isolation of Lac Repressor

The German biophysicist Adolf Wacker and his collaborators at Frankfurt University tried to set up an *in vitro* test for Lac repressor. To do so they measured the rate of transcription of salmon sperm DNA with RNA polymerase from *E. coli*. They then added extracts of *E. coli* grown in the presence and in the absence of IPTG. Extracts of *E. coli* grown in the absence of IPTG inhibited transcription of salmon sperm DNA by 40 percent whereas extracts of *E. coli* grown in the presence of IPTG inhibited by only 15 percent. From this, the authors concluded that they produced an *in vitro* assay for Lac repressor (1)!

They must have felt uneasy, because two years later they published another approach (2,3). Here they first synthesized tritiated thiomethyl-galactoside (TMG). They then added the radioactive inducer to growing cultures of *E. coli*. They centrifuged the bacteria, washed and sonicated them. Then they put the extract on a DEAE column and eluted it with a salt gradient. Free TMG eluted with the first fractions. But at a higher salt concentration they also eluted another peak of labelled TMG. Was this the Lac repressor-TMG complex? It was stable enough to be rechromatographed. It was destroyed by trypsin, but not by DNase or RNase. Was it present in constitutive mutants? It was absent in extracts from strain ML35 ($I^-Z^+y^-$), but present in extracts from strains ML308 and ML309 (both $I^-Z^+Y^+$)! They argued that the latter two strains may produce mutant repressors which still bind inducer (I^{-d} mutants we would say today). I argue that they had produced an *in vitro* test for Lac permease. The use of I^- nonsense, I^s, or I^- deletion mutants might have helped them to answer the question, but they never tried. In those days, mutants were devilish material for most German biochemists.

In a third and last effort they asked whether RNA or DNA co-purified with this high molecular fraction of label carrying the inducer TMG (3). This seemed to be the case. So they wrote they had isolated Lac repressor. To confuse the reader, they introduced a new name for Lac repressor. They called it the *"acceptor"*. This was the end! One of the students who participated in these experiments, Rainer Flöhl, left experimental science and became the excellent science editor of the conservative *Frankfurter Allgemeine Zeitung*, the German newspaper which may be compared to the New York Times.

References

(1) Wacker, A., Träger, L. & Chandra, P.: Einfluß von Methyl-β-thiogalactose und Lactose auf den "Repressor" der DNA abhängigen RNS-Polymerase. Naturwissenschaften **52**, 134, 1965.

(2) Lodemann, E., Drahovsky, D., Flöhl, R. & Wacker, A.: Über die spezifische Bindung von TMG in *E. coli*. Zeitschrift f. Naturforschung **22b**, 301-306, 1967.

(3) Dellweg, H., Lodemann, E., Drahovsky, D. & Wacker, A.: Acceptor for thiomethyl-galactoside in *Escherichia coli*. Biochem.Biophys.Res.Communications **26**, 71-74, 1967.

2.7 Isolation of Lac Permease and Arg Repressor

The method Mark Ptashne used to isolate λ *CI* repressor (1) had been used before in an attempt to isolate Lac permease (2): Non-induced and induced *E. coli* cultures were labelled separately with either ^{3}H or ^{14}C labelled amino acids. Their extracts were mixed and searched for the presence of the products of the *lac* operon. They should appear as peaks of ^{14}C over ^{3}H. The peak consisting of β-galactosidase can be identified easily by its enzymatic activity. The same holds for the peak consisting of transacetylase. Alan Kolber and Wilfred Stein, who did the work, found a third peak in the soluble fraction on a DEAE-cellulose column. They reasoned it to be (part of) Lac permease. Controls with proper mutants would have given clear results. The permease-less mutant the authors had used, where the peak was indeed missing, was not defined. Was it a nonsense or a missense mutant? We do not know. In addition they did not characterize the fraction. The paper was duly forgotten (3).

How would the method work if one analysed a mixture of two extracts of *E. coli*, one carrying a missense mutation in the *argR* (arginine repressor) gene and the other carrying the *argR$^+$* wild type? First I have here to remind the reader that missense mutants are not generally degraded. Then one has to keep in mind that Arg repressor represses eight genes. At least six of the eight gene products are soluble proteins. One would expect that turning on these genes would dominate the situation, so that nothing could be said about the different repressors. Yet, Shigezo Udake from the Institute of Physical and Chemical Research in Yamato-machi (Japan) interpreted these peaks in differences of ^{3}H/^{14}C label to indicate the presence of Arg repressor (4)! He thanked Bernie Davies and Luigi Gorini for discussion and was able to publish his paper in Nature. What should one think about this?

References

(1) Ptashne, M.: Isolation of λ CI repressor. Proc.Natl.Acad.Sci. USA **57**, 306-313, 1967.

(2) Kolber, A. & Stein, W.D.: Identification of a component of a transport "carrier" system: isolation of the permease expression of the *lac* operon of *Escherichia coli*. Nature **209**, 691-694, 1966.

(3) My thesis adviser, Kurt Wallenfels, held similar ideas. He actually thought that β-galactosidase is identical with Lac permease. Together with Howard Rickenberg (USA) he got a NATO grant to study and prove this proposition. I was the postdoc who was supposed to do the experiments. Indeed I went to Howard Rickenberg and worked in his lab for two years, but I never did a single experiment to this end. I worked instead on the specificity of induction of the *lac* system and tried – in vain – to isolate Lac repressor. Fortunately both men were lenient with me, and fortunately, too, the NATO research grant officials were lenient with them.

(4) Udake, S.: Isolation of the arginine repressor in *Escherichia coli*. Nature **228**, 236-238, 1970.

2.8 Genetic Proof that *lac* Operator is Dyadic

In 1971 the sequence of *lac* operator was unknown. It was unclear how and when it would be determined. Thus, it was the optimal moment to ask the question whether mutations could be used by geneticists to propose a particular structure. It made sense to assume dyadic symmetry of the *lac* operator. The 5'-3' sequence of the left upper strand would then be identical with the 5'-3' sequence of the right lower strand. One monomer of Lac repressor would interact with each half of *lac* operator. If indeed the interaction was totally symmetrical, even in the presence of RNA polymerase, then a mutation in a particular base pair on the left side of the operator should have the same effect upon repression as the corresponding mutation on the right side of the operator.

At the University of Colorado in Denver the late Jack R. Sadler and his postdoc Temple E. Smith set out to solve this problem. So they isolated about one thousand *lac* O^c mutants (1). Why so many? The authors quote as the motto of their paper: "Anything not forbidden is compulsory". This sentence indeed signals the dangers of this non-rational behaviour. They tested the repression values of all these mutants and found that they fell into several classes. The classes were different enough not to overlap each other. If the supposition of ideal, dyadic symmetry was right, each class should consist of *two* different mutations, one on the left and another on the right side of *lac* operator. Sadler and Smith used genetic crosses to order the mutations. The results confirmed the proposition of a dyadic structure of *lac* operator (2). I was a referee of their manuscript. I could *not* follow the reasoning in their calculations. I still cannot. The calculations seem to lack logic. I noted this and suggested that the manuscript should be published if the authors insisted; in the end truth would emerge.

Indeed the truth did emerge four years later, when Walter Gilbert and his students sequenced eleven of these O^c mutants (3). They transformed them into defective λ phage, grew up 150 liter cultures and isolated the phages. The phage DNA was sheared down by sonication to a size of 1,000 bps and then treated with the restriction enzyme *HaeI*, which had just become available. With the help of Lac repressor a 200 bp fragment carrying *lac O1* and *O3* was isolated on nitrocellulose filters. Then CAP protein and RNA polymerase were added to transcribe the DNA piece. The sequence of the transcribed RNA was then determined.

The result was sobering. There was *no* correlation between the map which Sadler and Smith had produced and the actual position of their operator mutations. The calculations I could not understand had been the self-delusion of the authors. Furthermore the authors could not know that the real *lac* operator is rather asymmetric. A mutation on the left (promoter proximal) side has a much stronger effect than the corresponding mutation on the right (promoter distal) side. Sadler and Smith had used as a motto of their mapping paper a sentence taken from Albert Einstein (2): "Our experience hitherto justifies us in believing that nature is the realization of the simplest conceivable mathematical ideas." This is true. But reality is often more complicated than the simplest mathematics. So, after debunking the Sadler and Smith paper, Walter Gilbert and his collaborators came to the conclusion that the assumption of dyadic symmetry of the protein-DNA contacts of the *lac* repressor-operator complex is itself an illusion. "These symmetries may just be statistical accidents, the true interaction of *lac* repressor and DNA may not in any sense involve a symmetry axis of the protein" (4). Sadler and Temple Smith guessed right but were wrong in their experiments. Gilbert was right in his experiments but no longer saw the obvious. What a mess.

References

(1) Smith, T.F. & Sadler, J.R.: The nature of lactose constitutive mutations. J.Mol.Biol. **59**, 273-305, 1971.
(2) Sadler, J.R. & Smith, T.F.: Mapping of the lactose operator. J.Mol.Biol. **62**, 139-169, 1971.
(3) Gilbert, W., Gralla, J., Majors, J. & Maxam, A.: Lactose operator sequences and the action of *lac* repressor. In: Protein-ligand interactions. Ed. by H. Sund & G. Blauer. de Gruyter, Berlin, New York, 193-210, 1975.
(4) Gilbert, W., Majors, J. & Maxam, A.: How proteins recognize DNA sequences. In: Dahlem Workshop on Chromosomes. Allfrey, V., Bautz, E., McCarthy, B., Schimke, R. & Tissières, A. (eds.), Dahlem. 167-176, 1976.

2.9 The Core of Lac Repressor Binds to *lac* Operator

It is always tempting to take the other side of a question which has not been settled. So, the late Jack Sadler and his post doc Temple Smith argued that *weak*, negative dominant constitutives (I^{-d}) are as informative as strong I^{-d} mutations for defining the region of Lac repressor which interacts with *lac* operator (1). I recall that Adler *et al.* (2) had found that 237 of 247 *strong* I^{-d} mutants map at the extreme 5' end of the *lacI* gene coding for Lac repressor. The other ten I^{-d} mutants mapped rather close to them but somewhat deeper in the *I* gene. Adler *et al.* (2) therefore concluded that a N-terminal protrusion of about 50 residues of Lac repressor does the recognition and binding of *lac* operator (see pages 70-75).

Five years after Adler *et al.* (2), Sadler and Temple Smith published a paper on their I^{-d} mutants (1). They concluded that weak I^{-d} mutations are distributed all along the *I* gene and that they are functionally as informative as strong I^{-d} mutants. So they concluded that the core of Lac repressor is also involved in *lac* operator recognition. But one year later, Walter Gilbert and his post doc Ronald Ogata showed by methylation protection that N-terminal tryptic fragments of Lac repressor ("headpieces", fifty-one or fifty-nine residues in length) bind weakly but specifically to *lac* operator (3). The monomeric headpieces bind to *lac* operator three orders of magnitude weaker than wildtype Lac repressor. So they had to use milligram amounts of headpiece to demonstrate that the methylation protection pattern is similar to the one found with pure Lac repressor.

Here, Kathleen Matthews at Rice University entered the field. She relied on the interpretation of Sadler and Smith (2) when she interpreted the effect on repression of chemical substitutions of cystein SH groups positioned in the core of Lac repressor (4-6). In several cases, operator binding was reduced after substitution of residues positioned in the core: But did this prove direct interaction of the core of Lac repressor with *lac* operator or did it just indicate indirect changes in a three dimensional structure? Kathleen Matthews believed that it did indicate direct interaction of the core of Lac repressor with *lac* operator.

It was tempting for her to propose that Ogata and Gilbert (3) had observed an artefact. Suppose that some intact Lac repressor was left in the mass of N-terminal peptides? Thinking along this line, Kathleen Matthews decided to do the reverse experiment: isolate Lac repressor core after trypsin treatment and analyse its possible operator binding! The core seemed indeed to bind weakly but specifically to *lac* operator (7-10). This encouraged her to propose her model of Lac repressor-operator interaction (11). In her model, all *four* cores of tetrameric Lac repressor are involved in specific recognition of *one lac* operator. The headpieces are supposedly involved in some non-specific backbone binding. Two years after this model was published, Brian Matthews proposed that the recognition helix of the helix-turn-helix motif of λ *cro* (of which he had just determined the X-ray structure) does the specific recognition of λ operator and that Lac repressor carries in its N-terminus a similar helix-turn-helix motif (12).

This was a challenge. Kathleen Matthews now showed that the tryptic core of Lac repressor was able, at high concentrations, to yield a methylation protection pattern of *lac* operator which was similar to the pattern obtained with intact Lac repressor. In addition, at high concentrations the tryptic core behaved like intact Lac repressor in several other tests (9-15). Of course, the argument she had raised against the Ogata-Gilbert experiment could also be raised against her own experiments. Suppose some of the alleged pure tetrameric core molecules still carried one or two N-terminal headpieces? It would have been easy for her to do an unambiguous experiment with the product of DNA lacking the region coding for headpiece. She never did this experiment.

In 1991 the Ogata-Gilbert experiment was repeated by Ponzy Lu with a modified headpiece of Lac repressor obtained from a synthetic DNA construct lacking the DNA which codes for the core of Lac repressor (16). This headpiece worked both *in vitro* and *in vivo*. Finally, NMR analysis by Robert Kaptein showed unambiguously that Lac repressor headpiece recognizes and binds *lac* operator (17,18). Now, finally, bromodeoxyuridine in position 3 of *lac* operator reacted with tyrosine 17 of Lac repressor in the laboratory of Kathleen Matthews (19). After thirteen years of disorder, the small world of *lac* was back in order.

References

(1) Miva, J. Sadler, J.R. & Smith, T.F.: Characterisation of I^{-d} repressor mutations of the lactose operon. J.Mol.Biol. **117**, 843-868, 1977.

(2) Adler, K., Beyreuther, K., Fanning, E., Geisler, N., Gronenborn, B., Klemm, A., Müller-Hill, B., Pfahl, M. & Schmitz, A.: How lac repressor binds to DNA. Nature **237**, 322-327, 1972.

(3) Ogata, R. & Gilbert, W.: An amino-terminal fragment of *lac* repressor binds specifically to *lac* operator. Proc.Natl.Acad.Sci. USA **75**, 5851-5854, 1978.

(4) Burgum, A.A. & Matthews, K.S.: Lactose repressor protein modified with fluorescein mercuric acetate. J.Biol.Chem. **253**, 4279-4286, 1978.

(5) Manly, S.P. & Matthews, K.S.: Activity changes in *lac* repressor with cysteine oxydation. J.Biol.Chem. **253**, 4279-4286, 1978.

(6) Brown, R.D. & Matthews, K.S.: Chemical modification of the lactose repressor protein using N-substituted maleimides. J.Biol.Chem. **254**, 5128-5134, 1979.

(7) Friedman, B.E. & Matthews, K.: Inducer binding to *Lac* repressor: effects of poly(d(A-T)) and trypsin digestion. Biochim.Biophys.Res.Comm. **85**, 497-504, 1978.

(8) Matthews, K.: Tryptic core protein of lactose repressor binds operator DNA. J.Biol.Chem. **254**, 3348-3353, 1979.

(9) O'Gorman, R.B., Dunaway, M., Matthews, K.S.: DNA binding characteristics of lactose repressor and the trypsin-resistent core repressor. J.Biol.Chem. **255**, 10100-10106, 1980.

(10) Dunaway, M. & Matthews, K.S.: Hybrid tetramers of native and core lactose repressor protein. J.Biol.Chem. **255**, 10120-10127, 1980.

(11) Dunaway, M., Manly, S.P. & Matthews, K.S.: Model for lactose repressor protein and its interaction with ligands. Proc.Natl.Acad.Sci. USA **77**, 7181-7185, 1980.

(12) Matthews, B.W., Ohlendorf, D.H., Anderson, W.F. & Takeda, Y.: Structure of the DNA binding region of *lac* repressor inferred from its homology with *cro* repressor. Proc.Natl.Acad.Sci. USA **79**, 1428-1432, 1982.

(13) Manly, S.P., Bennett, G.N. & Matthews, K.S.: Perturbation of *lac* operator DNA modification by tryptic core protein from *lac* repressor. Proc.Natl.Acad.Sci. USA **80**, 6219-6223, 1983.

(14) Manly, S.P. & Matthews, K.S.: *lac* operator DNA modification in the presence of proteolytic fragments of the repressor protein. J.Mol.Biol. **179**, 315-333, 1984.

(15) Manly, S.P., Bennett, G.N. & Matthews, K.S.: Enzymatic digestion of operator DNA in the presence of the *lac* repressor tryptic core. J.Mol.Biol. **179**, 335-350, 1984.

(16) Khoury, A.M., Nick, H.S. & Lu, P.: *In vivo* interaction of *Escherichia coli lac* repressor N-terminal fragments with the *lac* operator. J.Mol.Biol. **219**, 623-634, 1991.

(17) Boelens, R., Scheek, R.M., van Boom, J.H. & Kaptein, R.: Complex of *lac* repressor headpiece with a 14 base pair *lac* operator fragment studied by two dimensional nuclear magnetic resonance. J.Mol.Biol. **193**, 213-216, 1987.

(18) Chuprina, V.P., Rullmann, J.A.C., Lamerichs, R.M.J.N., van Boom, J.H., Boelens. R. & Kaptein, R.: Structure of the complex of *lac* repressor headpiece and an 11 base-pair half-operator determined by nucleic magnetic resonance spectroscopy and restrained molecular dynamics. J.Mol.Biol. **234**, 446-462, 1993.

(19) Allen, T.D., Wick, K.L. & Matthews, K.S.: Identification of amino acids in *lac* repressor protein cross-linked to operator DNA specifically substituted with bromodeoxyuridine. J.Biol.Chem. **266**, 6113-6119, 1991.

2.10 Lac Repressor Uses β-Sheet to Recognize *lac* Operator

The Russian biophysicist Georgy Gursky proposed, first in 1975 in a Russian (1), then in 1976 in an esoteric Western journal (2), and finally in 1977 at an international conference in New York (3), an elegant code for protein-DNA recognition. An antiparallel β-sheet placed in the minor groove of β-DNA supposedly recognizes either G/C or A/T pairs. The recognition is indirect. The side chains do not directly touch the bases, they determine the exact positioning of the backbone which then specifies the specific interactions. He used as the relevant part of the β sheet involved in DNA-sequence recognition the sequence proposed to be the recognition helix of Lac repressor from residues 17 to residue 26 (4).

Dimshaw Patel (5) and Gerald Fasman (6) had already predicted that this region formed a beta sheet. Others came to the same conclusion. A list of these faulty predictions of secondary structures can be found in (7). Furthermore Fasman predicted that this beta sheet would lie in the major groove of DNA (6). Two years later in 1977, an American group predicted that a beta sheet of residues 17 to 24 fits into the minor groove of *lac* operator (8). The structure of the complex looks rather like the model of Gursky, but – of course – the Americans did not cite him.

This situation changed in 1982 when Brian Matthews showed that the DNA binding region of Lac repressor is homologuous to the DNA binding region of λ *cro*, *i.e.* both contain a helix-turn-helix motif (9). The lingering doubts about the alpha helix were dispelled by Robert Kaptein with the solution of the NMR structure of the complex of Lac repressor headpiece and a *lac* operator fragment (10). When everybody else abandoned the beta sheet, Gursky kept the faith. He now proposed that the helix-turn-helix motifs seen in X-ray or NMR were artefacts (11-13). According to him they folded as antiparallel beta-sheets. As proof for his hypothesis he presented the following evidence: he synthesized various peptides which had the sequence of the helix-turn-helix motif of λ *cro*. He analysed the CD spectra of these peptides *in vitro* in the absence and presence of λ operator DNA. He concluded that the peptides converted into β-sheets in the presence of the λ operator DNA. Furthermore, he showed that the peptides protected λ operator DNA against DNase action (11-13).

Gursky's model made specific predictions for specificity changes. We tested several of them in the *lac* case here in Cologne and none worked. The specificity changes we found (14) were not reconcilable with Gursky's model. Yet he stuck to his model. Max Planck's dictum, "Old ideas die with those who propose them" (15), may apply here.

References

(1) Gursky, G.V., Tumanyan, A.S., Zasedatelev, A.S., Zhuze, A.L., Grokhovsky, S.L. & Gottikh, B.P.: A code governing specific binding of regulatory proteins to DNA and structure of stereospecific sites of regulatory proteins. Molekularnaya Biologiya **9**, 635-651, 1975.

(2) Gursky, G.V., Tumanyan, V.G., Zasedatelev, A.S., Zhuze, A.L., Grokhovsky, S.L. & Gottikh, B.P.: A code controlling specific binding of regulatory proteins to DNA. Molecular Biology Reports **2**, 413-425, 1976.

(3) Gursky, G.V., Tumanyan, V.G., Zasedatelev, A.S., Zhuze, A.L., Grokhovsky, S.L. & Gottikh, B.P.: A code controlling specific binding of protein to double helical DNA and RNA. In: Nucleic acid-protein recognition. Ed. by H.J. Vogel. Academic Press, New York, San Francisco, London. 189-217, 1977.

(4) Adler, K., Beyreuther, K., Fanning, E., Geisler, N., Gronenborn, B., Klemm, A., Müller-Hill, B., Pfahl, M. & Schmitz, A.: How *lac* repressor binds to DNA. Nature **237**, 322-327, 1972.

(5) Patel, D.J.: The predicted secondary structure of the N-terminal sequence of the *Lac* repressor and proposed models for its complexation to the *Lac* operator. Biochemistry **14**, 1057-1059, 1975.

(6) Chou, P.Y., Adler, A.J. & Fasman, G.D.: Conformational prediction and circular dichroism studies on the *lac* repressor. J.Mol.Biol. **96**, 25-45, 1975.

(7) Bourgeois, S., Jernigan, R.L., Szu, S.C., Kabat, E.A. & Wu, T.T.: Composite predictions of secondary structures of *lac* repressor. Biopolymers **18**, 2625-2643, 1979.

(8) Church, G.M., Sussman, J.L. & Kim, S.H.: Secondary structural complementarity between DNA and proteins. Proc.Natl.Acad.Sci. USA **74**, 1458-1462, 1977.

(9) Matthews, B.W., Ohlendorf, D.H., Anderson, W.F. & Takeda, Y.: Structure of the DNA binding region of *lac* repressor inferred from the homology with *cro* repressor. Proc.Natl.Acad.Sci. USA **79**, 1428-1432, 1982.

(10) Boelens, R., Scheek, R.M., van Boom, J.H. & Kaptein, R.: Complex of *lac* repressor headpiece with a 14 base-pair *lac* operator fragment studied by two dimensional nuclear resonance. J.Mol.Biol. **193**, 213-216, 1987.

(11) Grokhovsky, S.L., Surovaja, A.N., Sidorova, N.J., Votavova, H., Sponar, J., Frich, I. & Gursky, G.V.: Design and synthesis of peptides capable of binding specifically to DNA. Molekularnaya Biologiya **22**, 1315-1334, 1988.

(12) Grokhovsky, S.L., Surovaya, A.N., Sidorova, N.J. & Gursky, G.V.: Synthesis of nonlinear
     sequence-specific DNA binding peptide with specificity determinants similar to those of
     434 Cro repressor. Molekularnaya Biologiya **23**, 1558-1580, 1989.
(13) Grokhovsky, S.L., Surovaya, A.N., Brussov, R.V., Chernov, B.K., Sidorova, N.J. &
     Gursky, G.V.: Design and synthesis of sequence specific DNA-binding peptides. Journal
     of Biomolecular Structure & Dynamics **8**, 989, 1025, 1991.
(14) Lehming, N., Satorius, J., Kisters-Woike, B., Wilcken-Bergmann, B.v. & Müller-Hill, B.:
     Mutant Lac repressors with new specificities hint at rules for protein DNA-recognition.
     EMBO J. **9**, 615-621, 1990.
(15) Planck, M.: Persönliche Erinnerungen aus alten Zeiten. In: Vorträge und Erinnerungen.
     S. Hirzel Verlag, Stuttgart, 1949.

2.11 A π-Helix Binds to a Hairpin Loop of *lac* Operator

In 1966 Alfred Gierer proposed a special mode by which proteins like Lac repressor recognize their DNA operator targets: the operator DNA would not be present as a double-strand, but as two hairpin loops which could form when the operator has a dyad axis of symmetry (1). This proposition was discussed at length by Sobell for *lac* operator (2). Jones and Olson were more specific (3). They proposed that the N-terminal 25 residues of Lac repressor would form a continuous π-helix (4) which would interact with *lac* operator present as two hairpin loops. The basis for this proposal was the presence of tyrosine at positions 7, 12 and 17 of the N-terminus of Lac repressor. If these tyrosines intercalated into operator DNA, all three could lie only on the same side when the N-terminus was present as a π-helix. The π-helix was then and is still today a purely theoretical construct. On paper a π-helix may be constructed where just seven amino acids are needed for one turn. To the best of my knowledge a π-helix has never been observed in X-ray or NMR studies. However, there is no thorough theoretical work to demonstrate its nonexistence. In any case, the position of the O^c mutants (5) did not fit the model. The authors did not care. Just when their manuscript was submitted, a paper appeared in *Nature* which demonstrated that *lac* operator is *not* present as a hairpin (6). Were it present as a hairpin, it would be sensitive to certain DNases; but it is fully stable. So the model of Olson & Jones died before it was born. It had a nice obituary in a news-and-views article in *Nature* (7). If only such progress could be as rapid everywhere!

References

(1) Gierer, A.: Model for DNA and protein interactions and the function of the operator. Nature **212**, 1480-1481, 1966.
(2) Sobell, H.M.: Molecular mechanism for genetic recombination. Proc.Natl.Acad.Sci. USA **69**, 2483-2487, 1972.
(3) Jones, C.E. & Olson, M.O.: Sequence-specific DNA-protein interaction: the Lac repressor. J.theor.Biol. **64**, 323-332, 1977.
(4) Dickerson, R.E. & Geis, I.: The structure and action of proteins. Harper & Row, New York, 1969.

(5) Gilbert, W., Gralla, J., Majors, J. & Maxam, A.: Lactose operator sequences and the action of *lac* repressor. In: Protein-ligard interactions. Ed. by H. Sund & G. Blauer. de Gruyter, Berlin, New York, 193-210, 1975.

(6) Marians, K.J. & Wu, R.: Structure of the lactose operator. Nature **260**, 360-363, 1976.

(7) Malcolm, A.D.B.: Binding the *lac* repressor. Nature **268**, 196-197, 1977.

2.12 Prediction of the Sequence of *lac* Operator

The paper by Adler *et al.* (1) contains several propositions, some of which are correct and others which are wrong. I will list them here:

1. The N-terminal fifty residues of Lac repressor form a protrusion which recognizes *lac* operator (correct). This was based on solid genetic evidence (section 1.13.).
2. DNA recognition is made with an alpha helix which begins at residue 17 (correct). This was an educated guess.
3. *lac* operator consists of four identical or similar units (correct) joined head to tail (wrong). Four units seemed necessary for specific interaction. The text also mentions the possibility of a head to head connection with dyad symmetry. Indeed, two operators consisting of four half operators bind to one Lac repressor tetramer.
4. Specific interactions between the side chains of residues of the recognition helix with bases of the operator were predicted. The model turned out to be wrong. It was constructed with space filling models under the wrong assumption that as many side chains as possible would interact with bases. Thus, Gln26 and His29 were included to interact with bases. In reality neither interacts with a base of *lac* operator. This erroneous assumption misplaced all interactions.

The model building was done by me alone. I had nobody with whom to talk. The student revolution had hit my lab. There was chaos and nothing but chaos! The students told me that I had no right to give talks interpreting their data. My only postdoc told me I was exploiting him. Only one student, Alex Klemm, was actively interested in model building. The model he built of the peptide representing the alledged recognition helix was reversed in its N to C orientation. He refused to talk to me after I demonstrated this to him. He secretly wrote his own manuscript – using the data of the group – which he submitted to *Nature* without telling me or anybody else. Shortly thereafter he went to the Swiss alps for skiing. There he ridiculed the warning of an expert (*Hüttenwart*) that avalanches were expected. So he guided two postdocs of the Genetic Institute into the disaster. All three died in an avalanche. In retrospect I am astonished that I wrote this paper.

References

(1) Adler, K., Beyreuther, K., Fanning, E., Geisler, N., Gronenborn, B., Klemm, A., Müller-Hill, B., Pfahl, M. & Schmitz, A.: How *lac* repressor binds to DNA. Nature **237**, 322-327, 1972.

2.13 The Kinetics of *lac* Repressor-Operator Interaction

A new method or technique may open up a whole new area of research and theory. Such was the case when Arthur Riggs and Suzanne Bourgeois discovered the use of nitrocellulose filters to measure the kinetics and thermodynamics of the *lac* repressor-operator complex (1,2). Tetrameric Lac repressor binds, like many proteins, tightly to nitrocellulose. It still binds to nitrocellulose when one of its dimers binds *lac* operator. Double stranded DNA does not bind to nitrocellulose. By labeling the DNA one can thus measure the amount of *lac* repressor-operator complex (2,3). It took more than twenty years until the limitations of the test became apparent (4).

Dimeric active Lac repressor which binds to one copy of *lac* operator does not bind to nitrocellulose if its tail (residues 330-360) is cut off (5). It binds when some random sequences replace the two heptad repeats as is the case in the I^{adi} mutant (4). Tetrameric Lac repressor may bind with one set of its headpieces to *lac* operator and with the other set to nitrocellulose. Tetrameric Lac repressor may form a loop with DNA carrying two *lac* operators a suitable distance apart. But such a loop complex passes through the nitrocellulose filter. Loop formation may occur *in vitro* between the main operator *O1* and the auxiliary operator *O2* which are 401 bps apart (6). Lac repressor which only binds to *O1* or *O2* will stick to the filter. This has disturbed the interpretation of all measurements with *lac* DNA carrying *O1* and *O2*.

Recently we repeated these association rate measurements with dimeric and tetrameric Lac repressor (4). We used ideal *lac* operator embedded in DNA of various lengths lacking the auxiliary operators *O2* and *O3*. We found that the association rate was the same for dimeric and tetrameric Lac repressor and *lac* operator embedded in DNA 24, 84, 406 or 2,455 base pairs long. This argues against "sliding" of the Lac repressor along the DNA backbone, a mechanism often invoked (2, 3). Yet we found that tetrameric but not dimeric Lac repressor has its association rate increased tenfold when ideal *lac* operator is embedded in the 50,000 base pairs of λ DNA. How could one explain this? Possibly by "intersegment transfer" where tetrameric Lac repressor uses its free dimer to move around with increased speed within the mass of λ DNA.

To determine the dissociation rate of the *lac* repressor-operator complex, a hundredfold excess of cold *lac* operator DNA is added to the repressor-operator complex. This cold operator will form sandwich complexes, *i.e.* it will bind to the free dimer of the tetrameric *lac* repressor-operator complex. The sandwich complex passes through the filter like the loop complex. The extent of sandwich formation depends on the length of the cold *lac* operator fragment. The longer the fragment, the less sandwich complex is formed. This phenomenon was overlooked and misinterpreted as evidence for sliding (7).

References

(1) Riggs, A.D., Bourgeois, S., Newby, R.F. & Cohn, M.: DNA binding of the *lac* repressor. J.Mol.Biol. **34**, 365-368, 1968.

(2) Barkley, M.D. & Bourgeois, S.: Repressor recognition of operator and effectors. In: The operon. Ed. by J.H. Miller & W.S. Reznikoff. Cold Spring Harbor Laboratory, Cold Spring Harbor, NY, 177-220, 1978.

(3) Berg, O.G., Winter, R.B. & Hippel, P.H. v.: How do genome regulatory proteins locate their target sites? Trends in Biochem.Sciences **7**, 52-55, 1982.

(4) Fickert, R. & Müller-Hill, B.: How *lac* repressor finds *lac* operator *in vitro*. J.Mol.Biol. **226**, 59-68, 1992.

(5) Oehler, S., personal communication, 1994.

(6) Krämer, H., Niemöller, M., Amouyal, M., Revêt, B., Wilcken-Bergmann, B.v.: Lac repressor forms loops with linear DNA carrying two suitably spaced *lac* operators. EMBO J. **6**, 1481-1491, 1987.

(7) Whitson, P.A. & Matthews, M.S.: Dissociation of the lactose repressor-operator DNA complex: effects of size and sequence contexts of operator-containing DNA. Biochemistry **25**, 3845-3852, 1986.

2.14 Lac Repressor Bound to *O2* Acts as an Efficient Road Block

What happens to *lac* control if the two operators *O1* and *O3* are destroyed, and if only *O2* is left intact inside the *lacZ* gene? Will Lac repressor which is bound there act as an efficient "road block" and stop transcription? The answer is no. Even large amounts of Lac repressor, about one thousand molecules per cell, do not lead to any repression *in vivo* (1). A comparable experiment with synthetic DNA constructs carrying *Oid* downstream similarly showed no effect (2). But the literature comes to quite another conclusion. Based mainly on the work of Hermann Bujard and his colleagues (3), it is claimed that repression by road block is quite effective.

Bujard and his colleagues constructed a plasmid which carries a strong phage promoter which transcribes one message first, the DHFR (dihydrofolate reductase) gene and then the CAT (chloramphenicol acetyltransferase) gene. In between the DHFR and the CAT genes, a short DNA fragment was cloned which codes for 19 base pairs of *lac* promoter and the entire *lac O1* operator. In the presence of large amounts of Lac repressor no CAT is made. In contrast, in the presence of the inducer IPTG large amounts of CAT are made, as shown with a SDS PAGE gel. This suggests at least one hundredfold repression by a road block. The authors also measured the amounts of the respective mRNAs. And here comes the surprise. In the presence of Lac repressor they found about equal amounts of short DHFR message and of long DHFR-CAT message. In the presence of IPTG they found only the long DHFR-CAT message. This indicates about twofold, virtually no repression! The authors do not discuss this finding and take the SDS-protein gel as their only evidence. This is a case where we may never know the truth until somebody repeats Bujard's experiments.

It was observed a long time ago, but never explicitly stated, that the I^{Q1} promoter, which is about hundred times better than the normal I^+ promoter, does not increase β-galactosidase synthesis more than two fold. If there were unhindered read through through the *lac* promoter-operator control region one would expect a much higher increase. Deborah Steege analysed this phenomenon in detail. She showed convincingly that the *lac* promoter-operator region serves indeed as an efficient roadblock against transcription. Thus we conclude from her and our experiments that the presence of an intact promoter is essential for a roadblock. In its absence there is no roadblock. And Bujard's experiments? Are 19 base pairs of the *lac* promoter sufficient to create a roadblock? We do not know (4).

References

(1) Oehler, S., Eismann, E.R., Krämer, H. & Müller-Hill, B.: The three operators of the *lac* operon cooperate in repression. EMBO J. **9**, 973-979, 1990.

(2) Besse, M., Wilcken-Bergmann, B.v. & Müller-Hill, B.: Synthetic *lac* operator mediates repression through *lac* repressor introduced upstream and downstream from *lac* promoter. EMBO J. **5**, 1377-1381, 1986.

(3) Deuschle, U., Gentz, R. & Bujard, H.: *lac* repressor blocks transcribing RNA polymerase and terminates transcription. Proc.Natl.Acad.Sci. USA **83**, 4134-4137, 1986.

(4) Sellitti, M., Pavco, P.A. & Steege, D.: *lac* repressor blocks *in vivo* transcription of *lac* control region DNA. Proc.Natl.Acad.Sci. USA **84**, 3199-3203, 1987.

2.15 Auxiliary Operators Can be Disregarded in Repression

The wild type *lac* operon is about one thousandfold repressed. In 1988 it was not known that repression drops about seventyfold in the absence of the two auxiliary operators *O2* and *O3* (1). Thus Michael Lanzer and Hermann Bujard at Heidelberg University misinterpreted the reason for the decrease of repression they observed with various promoter constructs lacking the auxiliary operators and carrying just one *O1* operator (2). They thought that the low repression they observed in their constructs missing *O2* and *O3* was due exclusively to the high quality of their promoters. They overlooked the strong effects of the auxiliary operators. But of course the better RNA polymerase binds to its promoter the lower repression is with *O1*. RNA polymerase and Lac repressor compete for their presence on the promoter-operator DNA fragment (3). However the details of Bujard's interpretation and conclusion are in error for the reason mentioned.

References

(1) Oehler, S., Eismann, E.R., Krämer, H. & Müller-Hill, B.: The three operators of the *lac* operon cooperate in repression. EMBO J. **9**, 973-979, 1990.
(2) Lanzer, M. & Bujard, H.: Promoters largely determine the efficiency of repressor action. Proc.Natl.Acad.Sci. USA **85**, 8973-8977, 1988.
(3) Schlax, P.J., Capp, M.W. & Record, M.T.Jr.: Inhibition of transcription initiation by *lac* repressor. J.Mol.Biol. **245**, 331-350, 1995.

2.16 CAP Activates Transcription like λ *CI* Repressor

CAP and λ *CI* repressor both activate transcription. They both recognize their specific DNA binding sites with the help of a recognition helix which is part of a helix-turn-helix motif. Do they activate RNA polymerase by similar mechanisms? In λ *CI* repressor three residues were identified as involved in activation but not in DNA binding (1). Two residues are found at positions 2 and 6 of the preceding helix of the HTH motif. Both residues are acidic. To inactivate them they were replaced by either basic or neutral amino acids. These mutant λ *CI* repressors still bound to their DNA targets but they no longer activated transcription. So it was proposed that the two acidic residues activate RNA polymerase.

Is the same true for CAP? Nina Irwin, a student of Mark Ptashne, set out to answer this question (2). There is one acidic residue in the preceding helix of CAP but in a position different from the two acidic residues in λ repressor. Disregarding this discrepancy, she replaced this residue (E 171, in position 4 of the preceding helix) with lysine. She then tested this mutant CAP protein for activation *in vivo*, and for binding to the CAP DNA site *in vivo* and *in vitro*. Compared to wild type CAP her CAP mutant activated β-galactosidase synthesis thirty-fold less. *In vitro* this mutant seemed to bind to CAP DNA one third as well as wild type. Thus the decrease in activation could have been interpreted as being due to the failure of CAP binding to its target. However, she interpreted her data as having shown that glutamic acid 171 is indeed involved in activation. Several years later, three groups showed that the data were indeed incorrectly interpreted (3-5). If enough mutant CAP protein is produced to compensate for its lowered binding to the CAP site, activation is almost normal. In addition doubts have been voiced concerning the interpretation of the λ *CI* repressor mutants (6).

References

(1) Hochschild, A., Irwin, N. & Ptashne , M.: Repressor structure and the mechanism of positive control. Cell **32**, 319-325, 1983.
(2) Irwin, N. & Ptashne, M.: Mutants of the catabolite activator protein of *Escherichia coli* that are specifically deficient in the gene-activation function. Proc.Natl.Acad.Sci.USA **84**, 8315-8319, 1987.

(3) Bell, A., Gaston, K. Williams, R., Chapman, K., Kolb, A., Buc, H., Minchin, S., Williams, J. & Busby, S.: Mutations that alter the ability of *Escherichia coli* cyclic AMP receptor protein to activate transcription. Nucl.Acids Res. **18**, 7243-7250, 1990.

(4) Zhang, X., Zhou, Y., Ebright, Y.W., & Ebright, R.H.: Catabolite gene activator protein (CAP) is not an "acidic activating region" transcription activator protein. J.Biol.Chem. **267**, 8136-8139, 1992.

(5) Breul, A., Assmann, H. Golz, R., Wilcken-Bergmann, B.v. & Müller-Hill, B.: Mutants with substitutions for Glu 171 in the catabolite activator protein (CAP) of *Escherichia coli* activate transcription from the *lac* promoter. Mol.Gen.Genet. **238**, 155-160, 1993.

(6) Kolkhof, P. & Müller-Hill, B.: λ CI repressor mutants altered in transcriptional activation. J.Mol.Biol. **242**, 23-36, 1994.

2.17 In the Pitfall of Symmetry

By 1981 the binding site of the CAP protein upstream of the *lac* promoter had been sequenced. Several other similar CAP binding sites were also known. It was easy to see dyadic symmetry in the *lac* CAP site (1),

```
      <-<-<-<-<-   <- | ->   ->->->->->
5' T T A A T G T G A G T T | A G C T C A C T C T C T C A T T 5'
3' A A T T A C A C T C A A | T C G A G T G A G A G A G T A A 3'
      <-      <- | ->      ->
            <-<-<-<-            <- | ->          ->->->
```

but in the *gal* CAP site only one such half site could be seen. An optimistic inspection of the *gal* and *lac* half sites seemed to indicate that they were symmetrical too (1). Thus, Benoît de Crombrugghe and his collaborators proposed that *two* dimers of the CAP protein bound to the CAP site of *lac*, and that one CAP dimer bound to the CAP half site of the *gal* system. The X-ray structure of the complex of CAP protein and the CAP DNA site (2) argues against this proposition. No other study supported it. It was forgotten.

References

(1) O'Neil, M.C., Amass, K. & Crombrugghe, B.de: Molecular model of the DNA interaction site for the cyclic AMP receptor protein. Proc.Natl.Acad.Sci. USA **78**, 2213-2217, 1981.

(2) Schulz, S.C., Shields, G.C. & Steitz, T.A.: Crystal structure of a CAP-DNA complex: the DNA is bent by 90°. Science **253**, 1001-1007, 1991.

2.18 The CAP-DNA Complex: Two Propositions

The solution of the crystal structure of CAP protein complexed with cAMP was pioneering work (1). The structure immediately raised the question: how does CAP protein recognize its target DNA? We have to bear in mind that the CAP protein is an activator of transcription. Does it possibly change the architecture of DNA? Tom Steitz tried model building with ordinary right-handed B-DNA. He came to the conclusion that it was "not possible for CAP protein to interact with right-handed DNA" (1). So, he tried left-handed DNA. To his surprise left-handed DNA "fit snugly" (1). If this were true, then CAP protein would activate by drastically reordering and opening the DNA. It took ten years to discover that this is not so (2). To the best of my knowledge the authors never discussed what was wrong when they proposed that CAP protein would not bind to right-handed B-DNA. I presume they did not anticipate the massive bending of DNA by CAP protein (2).

F.R. Salemme from the University of Arizona spent a sabbatical at Yale at the time McKay and Steitz proposed their model of the CAP-DNA complex. He had a different idea to explain the activation of RNA polymerase by the CAP protein (3). Suppose CAP could bridge adjacent loops of a DNA solenoidal coil. Then CAP would effect a local redistribution of DNA twist-strain energy, thus resulting in the formation of a left-handed "solenoidal loop". Such a loop could provide the ideal DNA conformation to RNA polymerase with which to start transcription. Again, to the best of my knowledge no evidence was ever produced to prove this proposition.

References

(1) McKay, D.B. & Steitz, T.A.: Structure of catabolite activator protein at 2.9 Å resolution suggests binding to left-handed B-DNA. Nature **290**, 744-749, 1981.
(2) Schulz, S.C., Shields, G.C. & Steitz, T.A.: Crystal structure of a CAP-DNA complex: the DNA is bent by 90°. Science **253**, 1001-1007, 1991.
(3) Salemme, F.R.: A model for catabolite activator protein binding to supercoil DNA. Proc.Natl.Acad.Sci. USA **79**, 5263-5257, 1982.

2.19 The Adenine of Cyclic AMP Binds to DNA

Does the adenine moiety of cAMP possibly interact with a specific A-T base pair in the minor groove in the CAP-cAMP-DNA complex? Apparently the adenine residue is not positioned inside a cleft of CAP protein, since compounds such as indoleacetate compete with cAMP for binding to CAP protein. The indole ring looks to the CAP cleft like the cyclic phosphate/furanose part of cAMP. Activation of CAP by the indole compounds depends on the chemical nature of their side chain. So, Richard Ebright and James R. Wong reasoned it may be the adenine or the indole side chain which interacts with an A-T pair of the DNA target. The distortion of this A-T pair may provide RNA polymerase with the possibility to become active. This sounded neat. However the X-ray structure of the cAMP-CAP-DNA complex indicated that the adenine of cAMP is several Ångströms away from the DNA.

References

(1) Ebright, R.H. & Wong, J.R.: Mechanism for transcriptional action of cyclic AMP in *Escherichia coli*: Entry into DNA to disrupt DNA secondary structure. Proc.Natl.Acad.Sci.USA **78**, 4011-4015, 1981.
(2) Schulz, S.C., Shields, G.C. & Steitz, T.A.: Crystal structure of a CAP-DNA complex: The DNA is bent by 90°. Science **253**, 1001-1007, 1991.

2.20 The Structure of β-Galactosidase

The sequence of β-galactosidase was published in 1978 (1). In the same year Leroy Hood published a paper in which he proposed that residues 1-379 of β-galactosidase are duplicated in residues 398-781 and that its C-terminus is homologous to dihydrofolate reductase of *E. coli*. The similarities were rather weak. The solution of the X-ray structure of β-galactosidase (3) gave the opportunity to check the claims. β-galactosidase consists of the α peptide and five domains. The second domain, residues 220-334, is identical with the fourth domain, residues 628-736. This corresponds nicely with the homology seen by Hood in residues 243-342 and 635-728. But the structure of the other domains differ drastically from each other. Thus, the other homologies were an illusion. Moreover, the supposed homology with dihydrofolate reductase begins around residue 829 in the middle of the fifth domain which reaches from residue 737 to residue 1023. The three-dimensional structure of dihydrofolate reductase is solved and shows no similarity to the proposed homologous region in β-galactosidase. The take home lesson of this story is: the observation of the similarity of region 220-334 and 628-736 was correct. The rest was an illusion.

References

(1) Fowler, A. & Zabin, I.: Amino acid sequence of β-galactosidase. J.Biol.Chem. **253**, 5521-5525, 1978.
(2) Hood, L., Fowler, A.V. & Zabin, I.: On the evolution of β-galactosidase. Proc.Natl.Acad.Sci. USA **75**, 113-116, 1978.
(3) Jacobson, R.H., Zhang, X.J., DuBose, R.F. & Matthews, B.W.: Three-dimensional structure of β-galactosidase from *E. coli*. Nature **369**, 761-766, 1994.

2.21 From Lamarck to Cairns

How do mutations arise in a particular gene? Do they arise spontaneously at random, irrespective of the environment in which specific mutations are necessary for survival of an organism? Or do they occur more frequently in those genes where they are needed for survival than in others where they are not so needed? Does particular need trigger particular, useful mutations? Max Delbrück and Salvadore Luria gave a first, classical experimental answer to this question (1). They studied the frequency of the Ton A, B mutations which make E. coli resistant to phage T1. If mutations to resistance were induced by the presence of the phage, one would observe a Poisson distribution of their frequency in independent cultures. If they occurred independently of the presence of the phage, one would observe large fluctuations, i.e. jackpots. Indeed the latter was what Luria and Delbrück observed.

This is precisely what one would expect if one considers the mechanism of phage resistance. When E. coli cells are confronted with phage T1 they have no chance, i.e. no time to respond to this confrontation with "resistance". The phage attaches to a particular protein on the outer membrane of E. coli, injects its DNA and thereby kills the cell. Only those cells which lack this particular protein are resistant. Thus, the mutation which destroys this particular protein has to occur several generations before the phage comes close to the E. coli cell and attaches. The proteins to which the phage attaches are stable and have to be diluted out in consecutive rounds of cell duplication. Thus, the Luria-Delbrück experiment does not answer the general question it seems to ask. Delbrück was well aware of this. So he wrote in 1946: "In view of our ignorance of the causes and mechanisms of mutations, one should keep in mind the possible occurrence of specifically induced adaptive mutations" (2), i.e. he left the door open to Lamarckism.

Bacteria were for a long time the last stronghold of Lamarckism. I mentioned before that Lamarckism became a political issue when in 1948 it was declared the only allowed theory of inheritance in the Soviet Union (3). The few Western scientists who were not convinced by the Luria-Delbrück experiment were easily convinced by an elegant experiment of the Lederbergs (4) that Lamarckism was actually dead. The Lederbergs showed convincingly that resistance against the antibiotic streptomycin (SmR) occurs before the bacteria have come in contact with streptomycin. But again this is expected. The bacteria need several rounds of cell replication to dilute out the wildtype ribosomal protein which confers sensitivity against this antibiotic.

Thus, the question seemed solved and almost nobody bothered until in 1988 John Cairns pointed out that all the old experiments were ill-designed and that a proper experiment would look something like this (5): Suppose you plate a lac^-Z^- point mutation on lactose. Do the bacteria possibly sense the presence of the lactose which they have to metabolize to survive? Do they direct their mutations to the lac^-Z^- gene? Indeed he claimed to have found evidence for such behaviour. The mutation rate to lac^+Z^+ increased with incubation time on lactose. Now, one could argue that starvation would lead to a general increase of the mutation rate.

As a control, Cairns used valine resistance (5). *E. coli* K12 wild type cells do not grow in the presence of valine (6). Valine in the growth medium is transported into the cells. There it inhibits enzymes involved in leucine and isoleucine synthesis. Valine resistant mutants occur at a rate of 5×10^{-7} (6). There are many different genes in which mutations to valine resistance may occur. Cairns found that the number of valine resistant mutants did not increase as the number of lac^+ revertants increased. One would expect this result if valine resistance operated like phage *T1* or streptomycin resistance. Yet Cairns claimed that valine resistance "is rapidly expressed" (5) in contrast to *T1* and Sm resistance.

But is valine resistance really "rapidly expressed", *i.e.* expressed within one generation? Or is it only expressed after several generations like *TonA, B* or *Sm* resistance? If the latter were true, one would expect exactly the result he found! Cairns cites no evidence for his claim, nor does he give any evidence. The literature shows that this question has not been explicitly answered. Mutations in many genes may lead to valine resistance (6). One paper seems to indicate that the majority of the mutations are transport mutations, defective in valine transport from the medium into the cell (7). Here I would expect slow expression of the mutant phenotype. This would require the diluting out of the wild type protein which makes the cells valine sensitive! If this were so, indeed the number of valine resistant mutants should not increase with time of starvation.

In a further experiment Cairns used a lac^- mutant isolated by James Shapiro (the Shapiro mutant) where the excision of a fragment of phage *mu* leads to the selected lac^+ phenotype. Again, he seemed to prove his proposition. Finally, Cairns discussed possible mechanisms which could explain the phenomenon. Several of the mechanisms mentioned could later be excluded in particular cases. Not one could be demonstrated.

Cairns's paper (5) led to a small avalanche of supplementary papers supposedly presenting more evidence for the Lamarckian case and a few others claiming to have disproved the interpretation of the experiments. The defenders of the

Lamarckian view coined several words or terms for their phenomenon. I quote: instructive, adaptive, Cairnsian mutations. This is not the place to discuss these papers in detail. Most of these papers are cited in a recent review by a defender of the Lamarckian interpretation (8). Its author writes: "Even in peer-reviewed articles, scientists have exhibited a surprising fervor, verging on the religious, in debating this issue". This seems to me misguided self pity. I just refer the reader to the critical reviews by Richard Lenski and John Mittler (9) and by Paul Sniedgowski (10).

Within the recent debate several papers merit mention: One by James Shapiro (11) and another by John Cairns (12) which appeared back to back and demonstrated convincingly that there is *no* evidence for "directed" mutations in the case of the Shapiro *lac⁻* mutant. Here the phenomenon of the directed mutation simply disappeared upon close inspection. Then a paper appeared in Science which showed that the mutation rate to *lac⁺* increases only upon starvation when the bacteria are *recAB* wt (13). So it occured to three groups (14-16) that the *lac* genes which reverted from *lac⁻* to *lac⁺* happened to be located on an episome. Suppose the frequent reversions could only happen on a functioning, transferable episome but not on the chromosome! The relevant experiments were easily done. And that's what it turned out to be. The mutation rate on the chromosome is a hundred fold lower during starvation than on the episome. The functioning episome speeds up the appearance of revertants.

This time all the original Cairnsian experiments (5) had been refuted or explained. So as bottom line one can state: There is no evidence for directed, adaptive, Cairnsian or Lamarckian mutations in *E. coli.*

References

(1) Luria, S.E. & Delbrück, M.: Mutations of bacteria from virus sensitivity to virus resis-
     tance. Genetics **28**, 491-511, 1943.
(2) Delbrück, M.: Heredity and variations in microorganisms. Cold Spring Harbor
     Symp.Quant.Biol. **11**, 114, 1946.
(3) The situation in biological science. Proceedings of the Lenin Academy of Agricultural
     Sciences of the USSR. Session July 31-August 7, 1948, verbatim report. Foreign Lan-
     guage Publishing House, Moscow, 1949.

(4) Lederberg, J. & Lederberg, E.: Replica plating and indirect selection of bacterial mutants. J.Bacteriol. **63**, 399-406, 1952.

(5) Cairns, J., Overbaugh, J. & Miller, S.: The origin of mutations. Nature **335**, 142-145, 1988.

(6) De Felice, M., Levinthal, M., Iaccarino, M. & Guradiola, J.: Growth inhibition as a consequence of antagonism between related amino acids: effects of valine in *Escherichia coli* K-12. Microbiol.Rev. **43**, 42-58, 1979.

(7) Davis, E.J., Blatt, J.M., Henderson, E.K., Whitacker, J.J. & Jackson, J.H.: Valine-sensitive acetohydroxy acid synthetase in *Escherichia coli* K-12: unique regulation modulated by multiple genetics sites. Mol.Gen.Genet. **156**, 239-249, 1977.

(8) Foster, P.: Adaptive mutations: The use of adversity. Ann.Rev.Microbiol. **47**, 467-504, 1993.

(9) Lenski, R. & Mittler, J.E.: The directed mutation controversy and Neo-Darwinism. Science **259**, 188-194, 1993.

(10) Sniedgowski, P.: The origin of adaptive mutants: random or nonrandom? J.Mol.Evol. **40**, 94-101, 1995.

(11) Maenhaut-Michel, G. & Shapiro, J.A.: The role of starvation and selective substrates in the emergence of *araB-lacZ* fusion clones. EMBO J. **13**, 5229-5239, 1994.

(12) Foster, P.L.& Cairns, J.: The occurance of heritable *Mu* excisions in starving cells of *Escherichia coli*. EMBO J. **13**, 5240-5244, 1994.

(13) Harris, R.S., Longerich, S. & Rosenberg, S.M.: Recombination in adaptive mutation. Science **264**, 258-260, 1994.

(14) Radicella, J.P., Park, P.U. & Fox, M.S.: Adaptive mutations in *Escherichia coli*: a role for conjugation. Science **268**, 418-420, 1995.

(15) Galitski, T. & Roth, J.R.: Evidence that F plasmid transfer replication underlies apparent adaptive mutation. Science **268**, 421-423, 1995.

(16) Foster, P.L. & Trimarchi, J.M.: Adaptive reversion of an episomal frameshift mutation in *Escherichia coli* requires conjugal function. Proc.Natl.Acad.Sci. USA **92**, 5487-5490, 1995.

2.22 DNA Sequence Analysis: Not in this Century!

In 1968 Erwin Chargaff, the great scientist and prophet of doom wrote the following: "A detailed determination of the nucleotide sequence of a DNA molecule is beyond our present means, nor is it likely to occur in the near future ... Even the smallest functional DNA varieties seen, those occurring in certain small phages, must contain something like 5,000 nucleotides in a row. We may, therefore, leave the task of reading the complete nucleotide sequence of a DNA to the 21st century which will, however, have other worries" (1). Ten years later, the DNA sequences of the bacteriophage *fd* (2) and of the animal virus *SV40* (3,4) were published. There is more then a little irony here since Chargaff himself had worked out the hydrazine reaction which was used to sequence C or T of DNA (for details see 5). What can be learned from this episode? Do not believe an old scientist when he says something can never be done! But be aware, the reverse is similarly not necessarily true either! Never say never, never say always!

References

(1) Chargaff, E.: What really is DNA? Remarks on the changing aspects of a scientific concept. In: Progress in Nucleic Acid Research and Molecular Biology. Ed. by J.N. Davidson & W.E. Cohn. Academic Press, New York & London, 297-333, 1968.
(2) Beck, E., Sommer, R., Auerswald, E.A., Kurz, Ch., Zink, B., Osterburg, G., Schaller, H., Sugimoto, H., Sugisaki, K., Okamoto, T. & Takanami, M.: Nucleotide sequence of bacteriophage fd DNA. Nucl.Acids Res. **5**, 4495-4503, 1978.
(3) Fiers, W., Contreras, R., Haegeman, C., Rogiers, R., Va De Voorde, A., Van Heuverswyn, H., Van Hereweghe, J., Volkert, G. & Ysebaert, M.: Complete nucleotide sequence of SV40 DNA. Nature **273**, 113-120, 1978.
(4) Reddy, V.B., Thimmappaya, B., Dhar, R., Subramanian, K.N., Zain, B.S., Pan, J., Ghosh, P.K., Celma, M.L. & Weissman, S.M.: The genome of simian virus 40. Science **200**, 494-502, 1978.
(5) Maxam, A.M.: Nucleotide Sequence of DNA. In: Methods of DNA and RNA sequencing. Ed. by S.H. Weisman, Praeger, New York, 113-164, 1983.

2.23 Protein Splicing

Combine
, mutants
to restore
f(r)

Complementation is a phenomenon observed between certain mutants of the *lacZ* gene of *E. coli* (1). For example, a mutant which carries an internal deletion of codons 21 to 41 of the *lacZ* gene is *lac⁻* and does not produce even a trace of active β-galactosidase. This mutant is unable to form an active tetramer, it forms an inactive dimer. A deletion which deletes all the Z gene downstream of codon 60 and thus expresses just the first sixty N-terminal residues of β-galactosidase is also completeley inactive. But one observes β-galactosidase activity if one expresses *in vivo* the two polypeptides together in *E. coli* or if one mixes their extracts *in vitro*. The short N-terminal peptide binds to the core fragment in a noncovalent manner and leads to the formation of an active tetramer. This reaction has been called alpha complementation. A similar reaction is found with C-terminal peptids and C-terminal deletions of β-galactosidase. This is called omega complementation. An inspection of the recently solved X-ray structure of β-galactosidase demonstrates the details of these interactions (2).

In the early seventies B.N. Apte, a postdoc working in David Zipser's laboratory at Cold Spring Harbor, studied the fragments of β-galactosidase produced by various Z⁻ mutants. To study their complementation he produced various Z⁻/Z⁻ heterozygotes. If one plates such a heterozygote on a lactose EMB (eosin-methylene-blue) indicator plate, the colonies are white. If complementation occurs they carry red caps after 24 hours of incubation. If recombination occurs they carry red caps after 48 hours. Experience is needed to interpret such colonies correctly. Therefore it is better to perform the test in a *RecA* background. Here no recombination can occur. If one observes in a *RecA* strain colonies with red caps these are due to complementation.

Apte interpreted complementation to occur in one particular *RecA⁺* heterozygote where it was unexpected. So Zipser asked him to repeat the same experiment with a *RecA* strain. Apte did this and obtained the same result. He then used SDS acrylamide gels to determine the size of the fragments. He found to his and Zipser's surprise a polypeptide of normal β-galactosidase size, *i.e.* 135,000 Dalton instead of the expected two peptides. He had used a *RecA* strain, which made recombination between the two Z⁻ genes impossible. What was going on? Apte repeated the experiment with twenty-two different heterozygotes to confirm the result. He used nine different chromosomal Z⁻ mutants (six of them were *RecA*) and five different Z⁻ mutants on episomes. Sixteen of the twenty two constructs

[handwritten annotation: instead of 2 diff small prot, got 1 normal one]

produced large amounts of wild type β-galactosidase. Apte and Zipser concluded that under appropriate conditions peptides are spliced (3).

Peptide splicing was a new mechanism. Did it occur as well in antibodies when the variable and constant parts were put together? Alfred Hershey ("Heaven is where an experiment works and works") communicated the manuscript to PNAS (3). The interest of many people rose. Nobody could repeat the results. Three years later the paper was retracted by the authors (4). What went wrong?

An inspection of the paper shows that it lacks a section "Materials and Methods". The authors do not describe in the result section how they constructed and tested the *RecA* strains. The detailed genotypes of the strains are not given. There is no description how the crosses for heterozygotes were performed and how the heterozygotes were tested. This suggests that the postdoc had no idea how to perform and interpret a bacterial cross when he began his work. If it had never dawned upon Apte that he did his crosses without proper controls, he was no scientist but possibly an artist. If he did realize during his experiments that the crosses were wrong, he was guilty of fraud. First you deceive yourself, then others. The Arts do not know this type of fraud, they only know various forms of plagiarism. Fraud is a priviledge of the scientist. Fraud implies intent. But how do you demonstrate that somebody intentionally misinterprets his data?

That somebody misinterprets his data when he describes a phenomenon does not exclude that the phenomenon exists. The first cases of real protein splicing were described in 1990. More cases have been described since. For a recent review see (5).

References

(1) Ullmann, A., Jacob, F. & Monod, J.: Characterization by *in vitro* complementation of a peptide corresponding to an operator-proximal segment of the β-galactosidase structural gene of *Escherichia coli*. J.Mol.Biol. **24**, 339-343, 1967.

(2) Jacobson, R.H., Zhang, X.J., DuBose, R.F. & Matthews, B.W.: Three-dimensional structure of β-galactosidase from *E. coli*. Nature **369**, 761-766, 1994.

(3) Apte, B.N. & Zipser, D.: *In vivo* splicing of protein: One continuous polypeptide from two independently functioning operons. Proc.Natl.Acad.Sci USA **70**, 2969-2973, 1973.

(4) Apte, B.N. & Zipser, D.: Author's statement on polypeptide splicing. Proc.Natl.Acad.Sci. USA **73**, 661, 1976.

(5) Cooper, A.A. & Stevensen, T.H.: Protein splicing: selfsplicing of genetically mobile elements at the protein level. TIBS **20**, 351-356, 1995.

2.24 *O1*-Repressor-*O2* Loops

It is so easy to make a mistake. A certain sloppiness is inherent in all experiments. Often the effect one is looking for is minute. When in 1967 I was looking for inhibition of *lac* transcription by the recently isolated Lac repressor, I thought I had observed a minute effect. I said so to Walter Gilbert and he passed this on to Sidney Brenner who just visited the Harvard Biolabs. So Sidney Brenner came to my lab and asked for details. I had just repeated the experiment and the effect was gone. "Sorry", I said. "I just repeated the experiment and I could not confirm it." This was embarrassing, but it was true.

In 1976/77 I had the idea that Lac repressor may form a loop between *O1* and *O2* to increase repression. I, thus, predicted that a deletion of *O2* would decrease *lac* repression drastically. I had isolated internal deletions of the *lacZ* gene, which according to their mapping either had or had not deleted *O2*. I gave these deletions to Ingrid Triesch, a graduate student, to test their transacetylase level and I predicted that those mutants in which *O2* is deleted should have an elevated level, those which still have *O2* should have a normally repressed level (the deletions had to be in phase: those out of phase would lead to polarity effects, *i.e.* decrease levels of permease and transacetylase; thus, only some of the deletions were useful). The test for transacetylase was straightforward. When I gave the *lacZ* deletion strains to the student I told her what I expected, high or low levels of transacetylase. It was just marvelous. All predictions were fulfilled. Wherever I predicted *O2* to be deleted, the level of transacetylase was semi-constitutive, *i.e.* high.

I was so excited. For the first time loops were considered to play a role in repression. I wrote a manuscript for the Proceedings of the National Academy of Sciences (1). François Jacob communicated it. And of course I sent preprints to many people including Walter Gilbert. The paper was in press when I received a phone call from him. He had sequenced one of the deletions I had used. According to our analysis, the level of transacetylase was high, so *O2* was deleted. But his sequence showed unambiguously that *O2* was not deleted. What was going on? I went to the student and asked to see her data. She had them at home. I asked her to bring them in immediately. She did. And I discovered that some of her measurements were missing or contrived. There was no time to check closely. I telephoned PNAS and withdrew the paper.

It then turned out that most of her measurements were wrong or contrived. The real level of transacetylase is low in all of the deletions! Now, of course, we understand the reason for this. The presence of *O3* keeps repression almost normal. At that time I was so struck by the student's accomplishment that I could not think about loops anymore. The human problem replaced the problem of science. Not too long after the event I had a visit from a well known American geneticist who told me: "You were crazy to retract the loop paper. It is essentially right. Who cares that the student cheated!" I am glad that I did not follow his advice. What happened to the student? I do not know. In those days there were no committees. She simply left the lab and did not return. Thus seven years later, looping was first successfully demonstrated by Robert Schleif in the *ara* system (2). It took two more years until looping could be demonstrated in an artificial *lac* system (3) and four more years until my lab solved the old problem (4).

References

(1) Müller-Hill, B., Hobson, A.C., Mieschendahl, M. & Triesch, I.: Two operators control the *lac* operon: *lac* repressor binds both. Proc.Natl.Acad.Sci.USA **74**, April 1977, retracted while in press.

(2) Dunn, T.M., Hahn, S., Ogden, S. & Schleif, R.F.: An operator at -280 base pairs that is required for repression of *araBAD* operon promoter: addition of DNA helical turns between the operator and promoter cyclically hinders repression. Proc.Natl.Acad.Sci. USA **81**, 5017-5020, 1984.

(3) Mossing, M.C. & Record, M.T.: Upstream operators enhance repression of the *lac* repressor. Science **233**, 889-892, 1986.

(4) Oehler, S., Eismann, E.R., Krämer, H. & Müller-Hill, B.: The three operators of the *lac* operon cooperate in repression. EMBO J. **9**, 973-979, 1990.

Part 3

The *lac* Operon, a Paradigm of Beauty and Efficiency

3.1 The New Perspective

If we look at today's textbooks of genetics (1), biochemistry (2) or molecular biology (3, 4) we find the *lac* operon in a prominent place. But it is presented as if we still lived in the late seventies or early eighties. The results of more recent work are not mentioned. They seem of no general interest. Clearly interest has shifted from prokaryotic to eukaryotic organisms. At first glance this seems justified. In the last fifteen years an enormous amount of knowledge about eukaryotic systems has been gained. In contrast, the techniques used to analyse the *lac* system or any similar system have been optimized, but not replaced by new ones. Such is the case for DNA-cloning, DNA-sequencing, DNA-synthesis, X-ray or NMR analysis. None of these techniques is essentially new.

Yet the point of view from which one today regards a system like the *lac* system is a new one. Today, almost a dozen *E. coli* repressors are known which are similar in sequence to Lac repressor (5). They repress other systems, *e.g. gal*, purine (*pur*), raffinose (*raf*), *etc*. Most of these repressors are homodimers. They do not form heterodimers with each other. How do they manage their specific dimerization? The operators they bind to have different DNA sequences. How do they recognize different DNA sequences? They all use at least two operators, one main operator and one or two auxiliary operators. How are the operators positioned? How do they work? When the full sequence of *E. coli* DNA is determined, one may anticipate that additional repressors will be found which will be similar in sequence to Lac repressor. And *E. coli* is not the only bacterium around. How are these systems arranged in other bacteria? So, if possible, one would like to know all *possible optimal solutions* for systems like the *lac* system.

What I said here about Lac repressor is, of course, equally true for CAP protein. CAP is used as a positive transcription factor by many systems in *E. coli*. What are its optimal uses? How do they function? And finally, how is the level of glucose sensed by other bacteria? Is it always via cAMP and CAP? Analogous questions may be asked about the enzyme β-galactosidase and the Lac permease pump.

Today, crystal structures exist of the complex of tetrameric Lac repressor core with inducer IPTG (6), of Lac repressor (7), of the complexes of Lac repressor with its operator and with the inducer IPTG (7), of the complex of the homologous Pur repressor (8), of Pur repressor bound to its corepressor and operator (9), of the complex of CAP with its specific target DNA (10) and of β-galactosidase (11). The NMR structure of the complex between Lac repressor headpiece and its operator target has also been solved (12,13). So what is missing? The X-ray structures of CAP-RNA-polymerase-promoter complex and of Lac permease!

Knowledge is accumulating about all possible productive structures of this little universe. This leads to the conclusion that any part of a living system one analyses deeper than all others will become paradigmatic: as time goes by one will learn that more and more other systems have similar structures and function similarly. This is so because evolution does not produce a large number of completely different solutions for a particular problem. What works survives; what does not work becomes extinct. Elegant solutions are used again and again. The paradigmatic systems have thus revealed their beauty and efficiency earlier than others. Such was, is and will be the case with the *lac* system of *E. coli*.

References

(1) Suzuki, D.T., Griffith, A.H., Miller, J.H., Lewontin, R.C.: An introduction to genetic analysis. W. Freeman and Company, New York. 4th. ed. 1993.
(2) Stryer, L.: Biochemistry. W. Freeman and Company, New York. 4th. ed. 1995.
(3) Alberts, B., Bray, D., Lewin, J., Roberts, K. & Watson, J.D.: Molecular biology of the cell. Garland Publications. New York, London. 3rd. ed. 1994.
(4) Lewin, B.: Genes V. Cell Press, Cambridge. Oxford University Press, Oxford. 5th. ed. 1994.
(5) Weickert, M. & Adhya, S.: A family of bacterial regulators homologous to Gal and Lac repressor. J.Biol.Chem. **267**, 15869-15874, 1992.
(6) Friedman, A.M., Fischmann, T.O. & Steitz, T.A.: Crystal structure of *lac* repressor core tetramer and its implications for DNA looping. Science **268**, 1721-1727, 1995.
(7) Lewis, M., Chang, G., Horton, N.C., Kercher, M.A., Pace, H.C., Schumacher, M.A., Brennan, R.G. & Lu, P.: Crystal structure of the *Escherichia coli* lactose operon repressor and its complexes with DNA and inducer. Science **271**, 1247-1254, 1996.
(8) Schumacher, M.A., Choi, K.Y., Lu, F., Zalkin, H. & Brennan, R.G.: Mechanism of corepressor-mediated specific DNA binding by the purine repressor. Cell **83**, 147-155, 1995.
(9) Schumacher, M.A., Choi, K.Y., Zalkin, H. & Brennan, R.G.: Crystal structure of LacI member, PurR, bound to DNA: Minor groove binding by α helices. Science **266**, 763-770, 1994.
(10) Schultz, S.C., Shields, G.C. & Steitz, T.A.: Crystal structure of a CAP-DNA complex: the DNA is bent by 90°. Science **253**, 1001-1007, 1991.

(11) Jacobson, R.H., Zhang, X.J., DuBose, R.F. & Matthews, B.W.: Three-dimensional struc-
    ture of β-galactosidase from *E. coli*. Nature **369**, 761-766, 1994.

(12) Chuprina, V.P., Rullmann, J.A.C., Lamerichs, R.M.J.N., van Boom, J.H., Boelens, R. &
    Kaptein, R.: Structure of the complex of *lac* repressor headpiece and an 11 base-pair half-
    operator determined by nuclear magnetic resonance spectroscopy and restrained molec-
    ular dynamics. J.Mol.Biol. **234**, 446-462, 1993.

(13) Slijper, M., Bonvin, A.M.J.J., Boelens, R. & Kaptein, R.: Refined structure of *lac* repres-
    sor headpiece (1-56) determined by realaxation matrix calculations from 2D and 3D NOE
    data: change of tertiary structure upon binding to the *lac* operator. J.Mol.Biol. **259**, 761-
    773, 1996.

3.2 Some Numbers and Concepts

Before I outline the present state of knowledge of the *lac* system, it is appropriate to recall some numbers. An *E. coli* cell is about 2 μm long and has a diameter of about 1 μm. Thus its volume is about 2×10^{-12} cm^3. With the knowledge of Avogadro's number (6.023×10^{23}), the concentration of one molecule per *E. coli* cell can be calculated. Its concentration is about 10^{-9} molar. The concentration of DNA counted as base pairs is high (10^{-2} M). It has been estimated that the effective *in vivo* DNA concentration is one order of magnitude lower than the chemical concentration (1). Were DNA linear, the *lac Z* gene with its approximately 3,000 bps would be about 1 μm long. A sphere whose volume is about that of an *E. coli* cell has a radius of 1 μm.

E. coli has a single chromosome consisting of about 4.7×10^6 base pairs. A DNA sequence should be 11 base pairs long ($4^{11} \sim 4 \times 10^6$) to occur statistically only once in the *E. coli* chromosome. A particular DNA binding motif can recognize at most five base pairs. Thus a protein dimer with one DNA binding motif per monomer can barely recognize a unique DNA sequence in *E. coli*. These are numbers one should know by heart.

One cell of *E. coli* contains about 3,000 molecules of RNA polymerase. The weak promoter of the *lac* system can be activated fifty fold by the CAP protein in the presence of cAMP. In its activated state it is about a thousand fold repressed by wildtype Lac repressor. Various explanations have been suggested for activation and repression. I favour the simple ones: CAP protein bound to a suitably placed CAP site enhances functional binding (*i.e.* open complex formation) of RNA polymerase to a weak promoter by binding specifically to RNA polymerase (2). In contrast, Lac repressor bound to a suitably placed *lac* operator sterically inhibits functional binding (open complex formation) of RNA polymerase to its promoter (3). Thus activation is due to an increase in concentration of the active RNA polymerase-promoter complex and repression is due to a decrease in concentration of the active RNA polymerase-promoter complex.

Let's have a look at a fully induced or – which is here the same – a constitutive *lac* system at 37° (4). About 1.7 copies of the *lac* operon will be present in an average *E. coli* cell which is in the process of DNA replication. Every 1.7 seconds, one transcript is initiated from each fully induced or constitutive *lac* promoter. Thus about 23 RNA polymerase molecules transcribe at the same time one *lacZ* (β-galactosidase) gene which is about 3,000 bps long. The RNA polymerases are spaced about 135 nucleotides apart. Each one transcribes about 80 bases per second. From the resulting mRNAs about 3,000 molecules of tetrameric β-galactosidase are produced in one *E. coli* cell. *T7* Phage RNA polymerase is able to transcribe DNA eight times faster than ordinary *E. coli* RNA polymerase. If one transcribes the *lacZ* gene with this *T7* polymerase the yield of β-galactosidase drops drastically. RNA not translated is hydrolysed (5,6).

On the average there are about ten molecules of tetrameric Lac repressor in each *E. coli* cell (7). Thus the concentration of Lac repressor is about 10^{-8}M. Let us assume that under the conditions prevailing in the *E. coli* cell the binding constant of Lac repressor to *lac* operator is about 10^9M. Tetrameric Lac repressor consists of two dimers, each of which can bind to a *lac* operator (8,9). If we assume that initiation of transcription happens only when *lac* operator is not occupied by Lac repressor, a binding constant of 10^9 M would leave about five percent of *lac* operator in a free state and thus about five percent of the promoter activity. Thus we may expect maximally a twenty fold repression of *lac* transcription. This is the repression we find in the absence of both auxiliary operators *O2* and *O3* (10).

The *E. coli* cell consists of about one third solid material and two thirds water. Water is thus a rare solvent present in only a very few layers around the 3,000 different proteins of the *E. coli*. One has to consider water as part of the bacterial structure. To the uninitiated it may come as a surprise that 130 water molecules are released when Gal repressor binds to its operator (11).

References

(1) Hildebrandt, E.R. & Cozzarelli, N.R.: Comparison of recombination *in vitro* and in *E. coli* cells: measure of the effective concentration of DNA *in vivo*. Cell **81**, 331-340, 1995.

(2) Tagani, H. & Aiba, H.: Role of CRP in transcription activation at *Escherichia coli lac* promoter: CRP is dispensable after the formation of open complex. Nucl.Acids Res. **23**, 599-605, 1995.

(3) Schlax, P.J., Capp, M.W. & Record, T.M.: Inhibition of transcription initiation by *lac* repressor. J.Mol.Biol. **245**, 331-350, 1995.

(4) Kennell, D. & Riezman, H.: Transcription and translation initiation of the *Escherichia coli lac* operon. J.Mol.Biol. **114**, 1-21, 1977.

(5) Iost, I. & Dreyfus, M.: mRNAs can be stabilized by DEAD-box protein. Nature **372**, 193-196, 1994.

(6) Iost, I. & Dreyfus, M.: The stability of *Escherichia coli lacZ* mRNA depends upon the simultaneity of its synthesis and translation. EMBO J. **14**, 3252-3261, 1995.

(7) Gilbert, W. & Müller-Hill, B.: Isolation of lac repressor. Proc.Natl.Acad.Sci. USA **56**, 1891-1898, 1966.

(8) Kania, J. & Brown, D.T.: The functional repressor parts of a tetrameric *lac* repressor-β-galactosidase chimaera are organized as dimers. Proc.Natl.Acad.Sci. USA **73**, 3529-3533, 1976.

(9) Kania, J. & Müller-Hill, B.: Construction, isolation and implications of repressor-β-galactosidase hybrid molecules. Eur.J.Biochem. **79**, 381-386, 1977.

(10) Oehler, S., Eismann, E.R., Krämer, H. & Müller-Hill, B.: The three operators of the *lac* operon cooperate in repression. EMBO J. **9**, 973-979, 1990.

(11) Garner, M. & Rau, D.C.: Water release associated with specific binding of *gal* repressor. EMBO J. **14**, 1257-1263, 1995.

3.3 Activation of the *lac* Promoter by CAP Protein

3.3.1 The Modular Structure of CAP Protein

In 1982 the CAP gene was cloned and sequenced by a Japanese group and a French group at the Institut Pasteur (1,2). It codes for a protein composed of 209 residues. At physiological concentrations, CAP is present as a dimer. Each monomer has one binding site for cAMP. It has been shown with gel shift experiments (3) that a dimer devoid of cAMP does not bind specifically to its target DNA. A dimer of which only one CAP subunit binds cAMP but whose other subunit is free binds best to its specific target DNA. At higher concentrations of cAMP, when the two subunits each bind one molecule of cAMP, specific DNA binding decreases at least ten fold (3).

The crystal structure of the CAP dimer complexed with two molecules of cAMP was solved in 1981 and refined in 1982 and 1987 (4). In 1982 its structure could be compared to the CAP sequence (5) and to the structure of Cro repressor (7). Cro and CAP have helix-turn-helix (HTH) DNA binding motifs. The recognition helices (the C-terminal helices of the HTH motives) of Cro repressor dimer have the proper distance (34 Å) and the proper orientation to fit into two successive major grooves of ordinary B-DNA. The recognition helices of CAP also have the proper distance but they are oriented in such a way that they cannot fit into the major grooves of ordinary B-DNA (Fig. 8).

Ten years after the first CAP structure, Tom Steitz with his collaborators (8) succeeded in solving the structure of a CAP dimer complexed with two molecules of cAMP and a 30 bp DNA fragment carrying the consensus CAP binding site noted below (see also Fig. 9).

```
5'G C G A A A A G T G T G A C A T   A T G T C A C A C T T T T C G
   G C T T T T C A C A C T G T A   T A C A G T G T G A A A A G C G
   15          10 9 8 7 6 5 4 3 2 1   1 2 3 4 5 6 7 8 9 10       15
```

The DNA is kinked on both sides between T6 and G5! These kinks allow the side chains of residues 1, 2 and 6 of the recognition helix of both subunits to make specific contacts with bases 5 and 7. The DNA is thus bent by 90°. Earlier, in 1984, Steitz predicted moderate DNA-bending (9). Don Crothers then designed a quick and simple test to demonstrate protein induced DNA bending (10). He cloned a CAP binding site between two identical DNA repeats. The repeats were about one hundred base pairs long and full of restriction sites. If one now treats such a

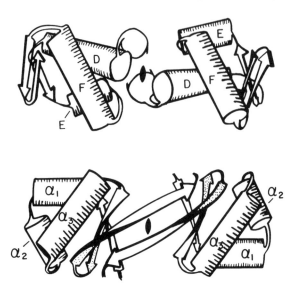

Fig. 8.: Schematic drawing comparing the backbone conformations in the DNA binding domains of CAP (upper) and *cro* (lower). F is the recognition helix of *cro* and a3 is the
recognition helix of CAP. From Steitz *et al.*, 1982 (7), Fig.1, with permission.

piece of DNA with one restriction enzyme after the other, one obtains DNA fragments of identical length which carry the CAP site at different positions. In a gel
shift experiment, the DNA fragments which have CAP protein bound to the CAP
site at their center move slower than those which carry it bound to sites at their
ends. This indicates bending of the DNA by CAP.

A recalculation of the CAP X-ray data indicated strong bending (11). But
how is bending in DNA induced? A novel idea and an interesting experiment
have recently been presented (12). Suppose bending occurs by asymmetric phosphate neutralisation? Then synthetic DNA carrying phosphate analogues lacking
charge should be bent, and indeed it does.

Let us now have a closer look at the modular structure of CAP. Figure 10 shows
the three-dimensional structure of one subunit of CAP. It consists of two domains.
One begins at its N-terminus and ends between helices C and D near residue 140.
There are proteases such as subtilisin which in the presence of cAMP cut only in
this region and nowhere else (3). The N-terminal peptide dimerizes via the C-
helix. It binds cAMP. The N-terminal peptide is similar in sequence to the regulatory cAMP binding subunit of cAMP-dependent protein kinase from bovine

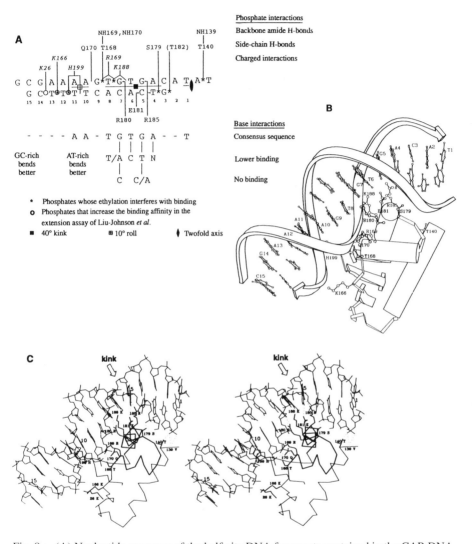

Fig. 9.: (A) Nucleotide sequence of the half-site DNA fragments contained in the CAP-DNA complex and assignments for interactions with the protein. (B) Schematic drawing of potential interactions between one DNA half-site and one small domain of the CAP dimer. Protein helices are shown as tubes and ß-strands as arrows. (C) Stereo view of one small domain of CAP (shown as an α-carbon backbone) bound to one half of the CAP binding site. The HTH motif is highlighted in bold. For details see Fig. 3 A-C from Schultz *et al.* 1991 (8), with permission.

CAP secondary structure

Helix[a]	Residues	β sheet[b]	Residues	β sheet	Residues
α_A	10–17	β_1	18–25	β_7	81–89
α_B	99–107	β_2	26–35	β_8	90–97
α_C	111–134	β_3	39–43	β_9	156–158
α_D	139–153	β_4	44–52	β_{10}	163–165
α_E	168–176	β_5	58–66	β_{11}	196–198
α_F	180–193	β_6	67–71	β_{12}	202–204

[a] α helix = 80 residues.
[b] β sheet = 75 residues.

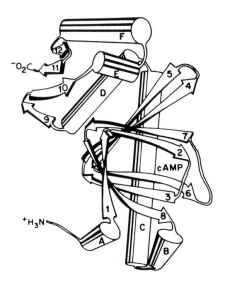

Fig. 10.: A schematic drawing of the CAP monomer. The regions of the polypeptide that are in α-helical conformation are represented by tubes and are lettered A to F. Those regions that are in ß-conformation are represented by arrows and numbered from 1 to 12. The approxiamte position of cAMP in the ß-roll is indicated. From McKay *et al.*, 1982, Fig. 3, (5) with permission.

heart muscle (13). The C-terminal domain which carries the HTH motif is similar in sequence and three-dimensional structure to a whole set of eukaryotic and prokaryotic protein binding domains such as LexA and BirA of *E. coli* and Histone H5 from chicken (14).

The reader who has come this far and who recalls Monod's work in diauxie with *B.subtilis* and *E. coli* may now exclaim: I finally understand! cAMP is used in eukaryotes and prokaryotes to up- or down-regulate genes! But now comes the embarrassing news. cAMP is not produced in *B.subtilis* (15,16) and in *E. coli* its level is the same in cells grown on glucose or lactose (17)! Glucose represses the *lac* system of *E. coli* by prohibiting the entry of lactose through an occasionally present Lac permease molecule. In the presence of glucose there is never enough lactose inside the uninduced cell to be transformed by the very few molecules of β-galactosidase into the true inducer allolactose.

If Monod and the rest of the world had continued with *B.subtilis*, completely disregarding *E. coli*, they would have ended with a close relative of Lac repressor. In *B.subtilis* the glucose effect is executed by phosphorylation and dephosphorylation of HPr (a phosphocarrier protein of the phosphotransferase system). And now comes the joke: phosphorylated HPr binds to a relative of Lac repressor, the *CcpA* gene product, which represses transcription of the glucose sensitive genes. Many roads lead to Lac repressor!

References

(1) Aiba, H., Fujimoto, S. & Ozaki, N.: Molecular cloning and nucleotide sequencing of the gene for *E. coli* cAMP receptor protein. Nucl.Acids Res. **10**, 1345-1361, 1982.

(2) Cossart, P. & Gicquel-Sanzey, B.: Cloning and sequence of the *crp* gene of *Escherichia coli* K12. Nucl.Acids Res. **10**, 1363-1378, 1982.

(3) Heyduk, T. & Lee, J.C.: *Escherichia coli* cAMP receptor protein: evidence for three protein conformational states with different promoter binding affinities. Biochemistry **28**, 6914-6924, 1989.

(4) McKay, D.B. & Steitz, T.A.: Structure of catabolite gene activator protein at 2.9 Å resolution suggests binding to left-handed B-DNA. Nature **290**, 744-749, 1981.

(5) McKay, D.B., Weber, I.T. & Steitz, T.A.: Structure of catabolite activator protein at 2.9 Å resolution. Incorporation of amino acid sequence and interactions with cyclic AMP. J.Biol.Chem. **257**, 9518-9524, 1982.

(6) Weber, I.T. & Steitz, T.A.: Structure of a complex of catabolite activator protein and cyclic AMP refined at 2.5 Å resolution. J.Mol.Biol. **198**, 311, 236, 1987.

(7) Steitz, T.A., Ohlendorf, D.H., McKay, D.B., Andersen, W.F. & Matthews, B.W.: Structural similarity in the DNA-binding domains of catabolite gene activator and *cro* repressor proteins. Proc.Natl.Acad.Sci. USA **79**, 3097-3100, 1982.

(8) Schultz, S.C., Shields, G.C. & Steitz, T.A.: Crystal structure of a CAP-DNA complex: the DNA is bent by 90°. Science **253**, 1001-1007, 1991.

(9) Weber, I.T. & Steitz, T.A.: Model of a specific complex between catabolite gene activator protein and B-DNA suggested electrostatic complementarity. Proc.Natl.Acad.Sci. USA **81**, 3973-3977, 1984.

(10) Liu-Johnson, H.-N., Gartenberg, M.R. & Crothers, D.M.: The DNA binding domain and bending angle of *E. coli* CAP protein. Cell **47**, 995-1005, 1986.

(11) Warwicker, J. Engelman, B.P. & Steitz, T.A.: Electrostatic calculations and model-building suggest that DNA bound to CAP is sharply bent. Proteins **2**, 283-289, 1987.

(12) Strauss, J.K. & Maher III, L.J.: DNA bending by asymmetric phosphate neutralisation. Science **266**, 1829-1834, 1994.

(13) Weber, I.T., Takio, K., Titani, K. & Steitz, T.A.: The cAMP-binding domains of the regulatory subunit of cAMP-dependent protein kinase and the catabolite gene activator protein are homologous. Proc.Natl.Acad.Sci. USA **79**, 7679-7683, 1982.

(14) Holm, L., Sander, C., Rüterjans, H., Schnarr, M., Fogh, R., Boelens, R. & Kaptein, R.: LexA repressor and iron uptake regulator from *Escherichia coli*: new members of the CAP-like DNA binding domain superfamily. Prot.Engen. **7**, 1449-1453, 1994.

(15) Henkin, T.M., Grundy, F.J., Nicholson, W.L. & Chambliss, G.H.: Catabolite repression of α-amylase gene expression in *Bacillus subtilis* involves a trans-acting gene product homologous to the *Escherichia coli lacI* and *galR* repressors. Mol.Microbiol. **5**, 575-584, 1991.

(16) Deutscher, J., Küster, E., Bergstedt, U., Charrier, V. & Hillen, W.: Protein kinase-dependent HPr/CcpA interaction links glycolytic activity to carbon catabolite repression in Gram-positive bacteria. Mol.Microbiol. **15**, 1049-1053, 1995.

(17) Inada, T., Kimata, K. & Aiba, H.: Mechanism responsible for glucose-lactose diauxie: reappraisal of cAMP dogma in *Escherichia coli*. Genes to Cells, **1**, 293-301, 1996.

3.3.2 Mutational Analysis of the *lac* Promoter and the CAP Site

In the *lac* system two promoters are present but *in vivo* only one is active (1-5). The active promoter (*P1*) is very weak with a suboptimal -10 region. +1 is the position of the first base pair transcribed (see Fig. 11). *P2*, the other promoter which is positioned 22 bp upstream of *P1*, is occupied by RNA polymerase in the absence of CAP. But RNA polymerase does not start from there.

In the absence of CAP-protein promoter *P1* is transcribed about fifty times less efficiently than in its presence. A double mutation (called UV5) in the -10 region of *P1* increases the quality of *P1* about thirty fold and thus makes it virtually independent of CAP protein (6). The two proteins which regulate the activity of *lac* promoter (CAP protein and Lac repressor) are bound in close vicinity to the *lac* promoter (Fig. 11). Position *and* quality of these sites determine the extent of activation and repression of the promoter.

Both the CAP site and the *lac* operators have a dyad axis of symmetry. They bind to suitably assembled dimers of CAP (complexed with one molecule of

A

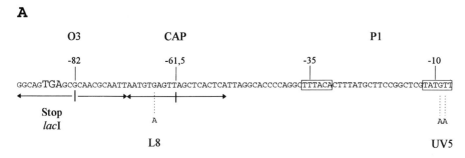

B

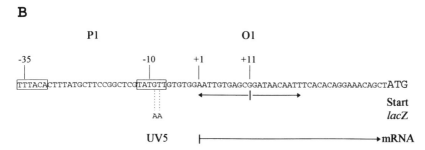

Fig. 11.: The 122 base pairs between the translation stop codon of the *lac I* gene and the translation start codon of the *lac Z* gene. The symmetry of *O3*, CAP and *O1* is indicated.

cAMP) or Lac repressor. The positions of the CAP site and the *lac* operators can be defined by the distance between the start site of transcription and the axes of symmetry of the CAP or *lac* operator sites.

The symmetry center of the CAP site is positioned 61.5 base pairs upstream of the main transcription start site of the *lac* promoter *P1* (we count the start base as half a base). Systematic analysis has shown that the position of the CAP site is always in one of the possible optima (*gal*, *mel*: 41.5 bp; *lac*: 61.5 bp; *malT*: 70.5 bp) (7-8). One subunit of CAP has to touch one α-subunit of RNA polymerase (see below!). It is noteworthy that evolution, like a designer, has been driven again and again to select optimal or near optimal positions for such binding sites.

The sequence of the CAP site of the *lac* system is itself suboptimal. An optimal, symmetric CAP site binds CAP protein about 450 fold tighter than the *lac* CAP site (10). How was the optimal, ideal binding site determined? Three methods can be used.

1. In the case of the CAP site where the sequences of many sites were known, their consensus sequence was deduced. The consensus sequence of all CAP sites is fully symmetrical. When the consensus CAP site was synthesized and tested, it was found to bind CAP protein about 450 fold tighter than the CAP site found in the *lac* system (10). It is a remarkable property of consensus sequences to come very, very close to ideal, *i.e.* best possible binding sequences.

2. A library of oligo DNAs can be synthesized and bound to the protein in question (11). Those DNAs which bind best elute latest and come closest to the ideal sequence. Optimal (ideal) *lac* operator was determined at a time when chemical DNA synthesis was not generally available. Here genomic mouse DNA was used as a library of quasi random DNA and screened for *lac* operator activity (12). A *lac* operator cloned on a multicopy plasmid titrates I^s repressor and makes *lac* I^s O^+ Z^+ cells constitutively *lac*$^+$. Such *lac* constitutive *lac*$^+$cells and plasmids can be selected.

3. A particular ideal sequence may be guessed and with chance the guess may be right. This succeeded once with *lac* operator (13) but is not recommended as a general procedure.

Most activator and repressor DNA binding sites are suboptimal. Their demands are rather different. If a CAP site is occupied to 50 percent by CAP, half the possible activation occurs. A further increase of the binding constant by a factor of 10 would increase activation from 50 to 90 percent, *i.e.* less than twofold. Clearly a further large increase in binding constant has only a small effect upon activation.

References

(1) Maquat, L.E. & Reznikoff, W.S.: *In vitro* analysis of the *Escherichia coli* RNA polymerase interaction with wild-type and mutant lactose promoters. J.Mol.Biol. **125**, 467-490, 1978.

(2) Malan, T.P. & McClure, W.R.: Dual promoter control of the *Escherichia coli* lactose operon. Cell **39**, 173-180, 1984.

(3) Meiklejohn, A.L. & Gralla, J.D.: Entry of RNA polymerase at the *lac* promoter. Cell **43**, 769-776, 1985.

(4) Peterson, M.L. & Reznikoff, W.S.: Properties of *lac* P2 *in vivo* and *in vitro:* an overlapping RNA polymerase binding site within the lactose promoter. J.Mol.Biol. **185**, 535-543, 1985.

(5) Donnelly, C.E. & Reznikoff, W.S.: Mutations in the *lac* P2 promoter. J.Bact. **169**, 1812-1817, 1987.

(6) Silverstone, A.E., Arditti, R.R. & Magasanik, B.: Catabolite-insensitive revertants of *Lac* promoter mutants. Proc.Natl.Acad.Sci.USA **66**, 773-779, 1970.

(7) Mandecki, W. & Caruthers, M.H.: Mutants of the *lac* promoter with large insertions and deletions between the CAP binding site and the -35 region. Gene **31**, 263-267, 1984.

(8) Gaston, K., Bell, A., Kolb, A., Buc, H. &. Busby, S.: Stringent spacing requirements for transcription activation by CRP. Cell **62**, 733-743, 1990.

(9) Ushida, C. & Aiba, H.: Helical phase dependent action of CRP: effect of the distance between the CRP site and the -35 region on promoter activity. Nucl.Acids Res. **18**, 6325-6330, 1990.

(10) Ebright, R.H., Ebright, Y.W. & Gunasekera, A.: Consensus DNA site for the *Escherichia coli* catabolite gene activator protein (CAP): CAP exhibits a 450-fold higher affinity for the consensus DNA site than for the *E. coli lac* DNA site. Nucl.Acids Res. **17**, 10295-10305, 1989.

(11) Perez-Fernandez, R., Arce, V. & Freedman, L.P.: Delineation of a DNA recognition element for the vitamin D3 receptor by binding site selection. Biochem.Biophys.Res.Comm. **192**, 728-737, 1993.

(12) Simons, A., Tils, D., Wilcken-Bergmann, B.v. & Müller-Hill, B.: Possible ideal *lac* operator: *Escherichia coli lac* operator-like sequences from eukaryotic genomes lack the central G.C. pair. Proc.Natl.Acad.Sci. USA **81**, 1624-1628, 1984.

(13) Sadler, J.R., Sasmor, H.B. & Betz, J.L.: A perfectly symmetric *lac* operator binds *lac* repressor very tightly. Proc.Natl.Acad.Sci. USA **80**, 6785-6789, 1983.

3.3.3 Mutational Analysis of the CAP Protein

In former times, *i.e.* in the sixties and seventies, mutant analysis meant the isolation of some random mutants, the description of their phenotypes and their mapping within a particular gene. At that time the structures of the proteins coded by the mutant genes were generally not known. Today things have changed radically. The sequence of a gene to be analysed is generally known. Often, the three-dimensional structure of the protein or its analogue is known. Such is the case here with the CAP protein (see section 3.3.1.).

The technique of making mutants has also changed completely. Today via site specific mutation one can change at will any codon at any location of any cloned gene. One can randomly mutagenize a small region of a cloned gene. Thus one may destroy particular regions in the modular structure of the protein encoded by it.

The most challenging approach today, suppression, has a long history in genetics. It implies the existence of two mutations. The first knocks out one activity. The second one (which is not simply a reversion of the first!) occurs elsewhere in the same gene or in another gene, and repairs the damage done by the first mutation. When separated from the first, the second mutation alone will also show some damage in the function of its gene (1).

To date three types of mutations of the CAP protein have been analysed:

1. Mutations which abolish the need of cAMP for activation.
2. Mutations which change the specificity of DNA target recognition.
3. Mutations which abolish activation without impairing DNA binding or bending.

3.3.3.1 Mutations of CAP which Relieve cAMP Requirement for Activation

Hiroji Aiba mutated a plasmid carrying the CAP gene with UV radiation, transformed *E. coli* cells unable to produce cAMP with this plasmid, and plated 10^5 transformants on lactose indicator plates (2). He obtained five mutants which in the absence of cAMP were *lac*$^+$. They mapped at codon position 53, 62, 141, 142 and 148 of CAP. At the same time Sankar Adhya in a equivalent experiment found similar mutants at codon positions 72, 141, 142 and 144 (3).

Residues 141 to 148 are part of the first α-helix of the DNA binding domain and residues 53 to 72 are part of two β-sheets one of which is close to this helix (Fig. 10). Although a detailed picture is lacking, these mutants help to explain the deformation of CAP brought by cAMP. The only weakness of these papers is that one would like to see at least ten times as many of such mutants in order to define all possible substitutions that lead to a similar phenotype.

Sankar Adhya inspected the crystal structure of CAP (see Fig. 9) and concluded that certain mutants of residues 141 and 148 (see above!) should be independent of cAMP. Indeed, charged amino acids in position 141 induce an cAMP independent phenotype, whereas lipophilic residues behave like wild type (4). Adhya also correctly predicted suppressor mutations in residues 138 which abolish the effect of the cAMP-independent mutants in residue 141. Finally substitutions were introduced at some of the residues which touch cAMP (5). The phenotypes of the resulting mutants were qualitatively as expected. Some needed a higher concentration of cAMP to become activated, others were almost inactive under all conditions.

3.3.3.2 Mutants of CAP which Change the Specificity of DNA Target Recognition

In 1984 the idea that the recognition helix of the helix-turn-helix (HTH) motif of CAP bound specifically to the CAP site was pure speculation. Richard Ebright, with the help of his colleagues at the Pasteur and Harvard, did the decisive experiment (6). They utilized a mutant, called L8, in the CAP site of the *lac* system. L8 lowers β-galactosidase expression about fifty fold. Such cells are phenotypically *lac⁻*. The L8 mutation had been sequenced. It carries a G to A substitution in position 5 of the CAP site.

```
10 9 8 7 6 5 4 3 2 1 () 1 2 3 4 5 6 7 8 9 10
5' A A T G T G A G T T () A G C T C A C T C A      wtCAP-lac
5' - - - - - A - - - - () - - - - - - - - - -      L8CAP-lac
```

CAP mutants had to be selected which would bind specifically to the mutant L8CAP site. They should have gained the lac⁺ phenotype. It was anticipated that they should have lost the capacity to bind to all other ordinary CAP sites. All other CAP systems would be inactive. Thus first CAP⁻ mutants were selected. Then, among these CAP- mutants, three were isolated which turned the *L8lac⁻* into lac⁺. I will not go into the details, I just note that the special knowledge of a generation of experts was used in this experiment. The three mutants were sequenced. They all carried substitutions in codon 181 of CAP. Glu was replaced with either Leu, Val or Lys. Codon 181 is residue two of the recognition helix! This *suppression* experiment shows convincingly that residue 181 of CAP binds directly to one of the bases of base pair 5 of the CAP site. This knowledge was immediately used for model building (7). For those who did not believe genetic data, two of the mutant CAP proteins were purified. They reacted *in vitro* with their DNA targets as predicted from the *in vivo* results (8).

So much for residue 181. But what about the other residues of the recognition helix? A suppression experiment is the best possible experiment but does not always work. So Ebright thought of a more general test, the loss-of-function test (9). Replace the residue you would like to analyse, say Arg 180 of CAP, with glycine: that is, remove the side chain. If you now test this glycine mutant with all possible base variants at the position where the original arginine interacted, it should not matter which base is now present. All bases at this particular position should function equally poorly, whereas one would observe differences in the other positions which are touched by other residues. The experiment indicated that Arg 180 reacts with base pair 5G/C. This was confirmed by the crystal structure (see Fig. 9). The final description of the various other base pair-amino acid interactions was given five years later (10).

3.3.3.3 Mutants of CAP which Abolish Activation without Impairing DNA Binding or Bending

Irrespective of the mechanisms of activation of transcription by CAP, mutants must exist which do not activate but which still bind to the target DNA: such mutants are most easy to explain if one assumes that the activator CAP activates by binding RNA polymerase, such that the local concentration of the latter increases at its poor promoters. Additionally when we assume that CAP induces some allosteric change in RNA polymerase, a region in CAP must exist which makes this interaction. When one proposes that CAP activates transcription by DNA-distortion, *i.e.* bending DNA close to the promoter, one would expect such mutants to bind to the CAP site but not to bend the DNA. And finally, even if one believes in a mechanism one does not yet quite understand, such mutants should exist. And so such mutants were looked for by three groups and found (11-12). Whenever these mutants were analysed in detail, bending was as normal as DNA binding. Thus DNA distortion, *i.e.* bending, is so intimately coupled to CAP-DNA binding that it could not be separated. The complex between CAP-site-promoter-DNA, CAP and RNA polymerase uses this bend. Occams razor thus demands that the function of the bend is to make this structure possible and only just that.

Mutants defective in activation were also found in FNR, a close relative of CAP. In FNR they map in the same positions as in the corresponding CAP mutants (13). We will now concentrate on the analysis done with the CAP mutants by Richard Ebright and his collaborators (14-16). They randomly mutagenized the CAP gene and isolated 21 CAP mutants which did not activate (14). Like the mutants mentioned above, they all occurred in a single region of CAP, extending from residue 156 to residue 162. Inspection of the crystal structure of CAP indicates that they occur in the loop between β-sheets 9 and 10 of the DNA binding domain (see Fig. 10). They then used targeted saturation mutagenesis of codons

152 to 166 of the CAP gene to isolate 200 mutants defective in transcription activation (15). All single substitutions mapped in codons 156, 157, 158, 159, 160, 162, 163 and 164 of CAP.

Is the effect of these mutants the same irrespective of where the CAP protein is positioned? Recall that in both the *gal* and *mel* systems it is positioned at -41.5 bp, in *lac* at -61.5 bp and in *malT* at -70.5 bp upstream of the start of transcription. A detailed analysis indicated that the activation mutants which were originally isolated in the *lac* system had qualitatively similar but quantitatively different effects at the position found in the *gal* system (16).

CAP placed with its center 41.5 bp upstream from the transcription start is very close to RNA polymerase. It has to touch it. So Ebright and his collaborators asked this question: is it actually the same subunit of the CAP dimer which touches RNA polymerase when placed at the distance of 41.5 bp or 61.5 bp, or are both subunits touched in both cases (17)?

The experiment can be done by simultaneously expressing one of the mutants defective in activating the *lac* or the *gal* promoter together with normally activating CAP which binds only to the mutated L8 CAP target (see section 3.3.). Mixed dimers of CAP will be formed. Asymmetric CAP sites, which have the normal wild type sequence on one side, and the mutated L8 sequence on the other side are then placed 41.5 bp and 61.5 bp upstream of a CAP dependent promoter. The four possible target systems are thereby brought into contact with the mixed dimers of CAP and the relevant effects are measured. The answer is clearcut (17): at the distance of 41.5 bp it is the promoter distal subunit of CAP which has to be intact in order to touch RNA polymerase. At the distance of 61.5 bp it is the promoter proximal subunit of CAP which has to be intact.

References

(1) I cannot resist telling a story about suppression. In 1969 I met a postdoc in Cold Spring Harbor who proposed that suppression will become the poor man's X-ray crystallography. One evening Jim Watson invited me for dinner. We had almost finished when the door opened and the postdoc entered. He was not invited. Jim asked him to leave. The postdoc tried to play it as if it were a misunderstanding. He took a glass and said hospitality was certainly a quality of Jim's. Jim threw him out. During the winter of 1972 I again met the postdoc now in David Suzuki's office in Vancouver. He explained that he

was travelling. He was on his way to the planets and from there to the milky way. To bring him down to earth I asked him about intracistronic suppression. Oh, he was not interested in such details anymore. May others exploit his idea. He was travelling to the stars. No structure was ever solved by suppression. It was all total nonsense! The path of unreason should be resisted right from the start. Otherwise it impedes and finally destroys. Jim was right that night as he so often has been before and after.

(2) Aiba, H., Nakamura, T., Mitani, H. & Mori, H.: Mutations that alter the allosteric nature of cAMP receptor protein. EMBO J. **4**, 3329-3332, 1985.

(3) Garges, S. & Adhya, S.: Sites of allosteric shift in the structure of cyclic AMP receptor protein. Cell **41**, 745-751, 1985.

(4) Kim, J., Adhya, S. & Garges, S.: Allosteric changes in the cAMP receptor protein of *Escherichia coli*: Hinge reorientation. Proc.Natl.Acad.Sci. USA **89**, 9700-9704, 1992.

(5) Lee, E.J., Glasgow, J., Leu, S.-F., Belduz, A.O. & Harman, J.G.: Mutagenesis of the cyclic AMP receptor protein of *Escherichia coli*: targeting positions 83, 127 ans 128 of the cyclic nucleotide binding pocket. Nucl.Acids Res. **22**, 2894-2901, 1994.

(6) Ebright, R.H., Cossart, P., Gicquel-Sanzey, B. & Beckwith, J.: Mutations that alter the DNA sequence specificity of catabolite gene activator protein of *E. coli*. Nature **311**, 232-235, 1984.

(7) Ebright, R.H., Cossart, P., Gicquel-Sanzey, B & Beckwith, J.: Molecular basis of DNA sequence recognition by the catabolite gene activator protein: detailed inferences from three mutations that alter DNA sequence specificity. Proc.Natl.Acad.Sci. USA **81**, 7274-7278, 1984.

(8) Ebright, R.H., Kolb, A., Buc, H., Kunkel, T.A., Krakow, J.S. & Beckwith, J.: Role of glutamic acid -181 in DNA-sequence recognition by the catabolite gene activator protein (CAP) of *Escherichia coli*: altered DNA-sequence-recognition properties of (Val[181])CAP and (Leu[181])CAP. Proc.Natl.Acad.Sci. USA **84**, 6083-6087, 1987.

(9) Zhang, X. & Ebright, R.H.: Identification of a contact between arginine -180 of the catabolite gene activator protein (CAP) and base pair 5 of the DNA site in the CAP-DNA complex. Proc.Natl.Acad.Sci. USA **87**, 4717-4721, 1990.

(10) Gunsakere, A., Ebright, Y.W. & Ebright R.H.: DNA sequence determinants for binding of the *Escherichia coli* catabolite gene activator protein. J.Biol.Chem. **267**, 14713-14720, 1992.

(11) Bell, A., Gaston, K., Williams, R., Chapman, K., Kolb, A., Buc, H., Minchin, S., Williams, J. & Busby, S.: Mutations that alter the ability of *Escherichia coli* cyclic AMP receptor to activate transcription. Nucl.Acids Res. **18**, 7243-7250, 1990.

(12) Eschenlauer, A.C. & Reznikoff, W.S.: *Escherichia coli* catabolite gene activator protein mutants defective in positive control of *lac* operon transcription. J.Bact. **173**, 5024-5029, 1991.

(13) Williams, R., Bell, A., Sims, G. & Busby, S.: The role of two surface exposed loops in transcription activation by the *Escherichia coli* CRP and FNR proteins. Nucl.Acids Res. **19**, 6705-6712, 1991.

(14) Zhou, Y., Zhang, X. & Ebright, R.H.: Identification of the activating region of catabolite gene activator protein (CAP): isolation and characterisation of the mutants of CAP specifically defective in transcription activation. Proc.Natl.Acad.Sci. USA **90**, 6081-6085, 1993.

(15) Niu, W., Zhou, Y, Dong, Q., Ebright, Y.W. & Ebright, R.H.: Characterisation of the activating region of *Escherichia coli* catabolite gene activator protein (CAP) I. Saturation and alanine scanning mutagenesis. J.Mol.Biol. **243**, 595-602, 1994.

(16) Zhou, Y., Merkel, T.J. & Ebright, R.H.: Characterisation of the activating region of *Escherichia coli* catabolite gene activator protein (CAP) II. Role of class I and class II CAP-dependent promoters. J.Mol.Biol. **243**, 603-610, 1994.

(17) Zhou, Y., Pendergast, P.S., Bell, A., Williams, R., Busby, S. & Ebright, R.H.: The functional subunit of a dimeric transcription activator depends on promoter architecture. EMBO J. **13**, 4549-4557, 1994.

3.3.4 RNA Polymerase, a Partner of CAP

The simplest explanation of the genetic experiments in which only the activation but not DNA binding and bending capacity of CAP is destroyed, proposes a direct interaction between CAP and RNA polymerase. One predicts that wild type CAP binds RNA polymerase under those conditions where such CAP mutants do not bind. Such an experiment was performed (1). CAP binds in the presence of cAMP to a fluorescent, synthetic, 42 bp long symmetric CAP site. The DNA contains no -35 polymerase site. RNA polymerase binds with a binding constant of 3×10^{-7} M to wild type CAP bound to such a CAP site, whereas it does not bind to mutant (*ala* 158) CAP.

The next question is: which subunit of RNA polymerase is contacted by CAP? Let me remind the reader that the *E. coli* RNA polymerase which interacts with CAP-dependent and CAP-independent promoters consists of five subunits: two α, one β, one β' and one σ. Kazuhio Igarashi and Akira Ishihama showed by *in vitro* transcription experiments that the α subunit is involved in CAP binding (2). RNA polymerase containing α subunits which lack the 73 C-terminal residues is active at a CAP independent promoter. In the presence of the *lac* promoter and CAP it is inactive, in contrast to wt polymerase. One year later, the same authors demonstrated that point mutations in codons from 265 to 270 of the α subunit show the same effect (3). Then Ebright and his colleagues showed that the α subunit of RNA polymerase has a modular structure (4). It consists of one stable N-terminal (8-241) and one stable C-terminal (249-329) domain. The C-terminal domain dimerizes and seems to bind specifically in the -40 to -60 region of CAP-independent promoters (5). At the moment, the experts are waiting for the isolation of a suppressor mutant in the CAP/α system. A particular CAP mutant which does not activate should be suppressed by a particular mutant in α. So far this mutant has not been found.

By now it is clear that CAP activates RNA polymerase by binding to it. Before discussing the mechanism by which CAP activates transcription, I have to present the main steps RNA polymerase takes in transcription. RNA polymerase first forms with its promoter a closed complex, in which DNA remains mainly in its B-conformation. This complex then isomerises to an open complex, in which the double strand is opened around the -10 box. Then the initiated complex is formed and transcription begins. It has been shown that in the presence of CAP the amount of closed complex is increased at least 13 fold, but that in the presence of CAP the rate of conversion of closed complex into open complex is not changed (6).

Thus, CAP simply increases the local concentration of RNA polymerase close to the ineffective promoter. But how can it increase the local concentration of RNA polymerase when a single *E. coli* cell contains about 3,000 molecules of RNA polymerase? One has to recall that most, about 99 percent, of these molecules are either active in transcribing DNA or irreversibly bound to active promoters. Thus, only a very low amount of free RNA polymerase is available. Its concentration can be sufficiently increased by appropriately bound activator which binds weakly to RNA polymerase. Thus, it is simply binding and not tickling as Mark Ptashne once proposed. Those who like Occam's razor need not invoke an allosteric change in RNA polymease.

It is interesting to note that a property common to many eukaryotic promoters is found with CAP as well. If two CAP molecules bind, one in position -41.5 and the other in -90.5, then they activate synergistically (7). The same is true when Lambda repressor and CAP bind in positions -41.5 and -93.5 respectively (8). It is strange that Lambda repressor supposedly binds to a particular residue of the σ subunit of RNA polymerase with a residue of the preceeding helix of the HTH motif (9). Possibly those who doubt this allocation (10) are not entirely wrong and Lambda repressor indeed interacts with RNA polymerase in a manner similar to CAP.

CAP is active only when it is bound rather closely to RNA polymerase and when it is presented on the proper side of DNA. It could therefore be predicted that CAP is inactive *in vitro*, when the CAP site is connected via four bp of single stranded DNA to the promoter (11). Under these conditions, when the DNA can move in all directions, CAP protein is unable to properly increase the local concentration of RNA polymerase. Furthermore, *in vitro* CAP is dispensable after the formation of the open complex (12). So CAP seems to operate in a manner similar to the operation of Gal11 in yeast cells, which only needs to contact a component of the polymerase II holo enzyme to activate transcription (13).

Finally, I return to the question of whether there is any functional role for the 90° DNA bent induced by CAP? The bend determines the structure of the CAP polymerase complex, but is that all? This question was raised when it was shown that the presence of bent DNA upstream makes the *lac* promoter independent of CAP (14,15). We know now that the α subunit of RNA polymerase may bind to these upstream regions and thereby increase the transcription rate (16). But it is reassuring that no evidence was found that the energy of the bend is used to initiate transcription (17). The search for such an effect was negative.

The expert will have certainly noticed that I did not quote from some papers in which conclusions are reached contrary to my interpretation of CAP activity. The curious non-expert may be interested in reaching his own conclusions. He should consult two reviews in which *all* relevant papers are quoted (18,19).

References

(1) Heyduck, T., Lee, J.C., Ebright, Y.W., Blatter, E.E., Zhou, Y. & Ebright, R.H.: CAP interacts with RNA polymerase in solution in the absence of promoter DNA. Nature **364**, 548-549, 1993.

(2) Igarashi, K. & Ishihama, A.: Bipartite functional map of the *E. coli* RNA polymerase α subunit: involvement of the C-terminal region in transcription activation by cAMP-CRP. Cell **65**, 1015-1022, 1991.

(3) Zou, C., Fujita, N., Igarashi, K. & Ishihama, A.: Mapping the cAMP receptor protein contact site on the α subunit of *Escherichia coli* RNA polymerase. Mol.Microbiol. **6**, 2599-2605, 1992.

(4) Blatter, E.E., Ross, W., Tang, H., Gourse, R.L. & Ebright, R.H.: Domain organisation of RNA polymerase α subunit: C-terminal 85 amino acids constitute a domain capable of dimerization and DNA binding. Cell **78**, 889-896, 1994.

(5) Ebright, R.H. & Busby, S.: The *Escherichia coli* RNA polymerase α subunit: structure and function. Curr.Op.in Gen.and Dev. **5**, 197-203, 1995.

(6) Malan, T.P., Kolb, A., Buc, H. & McClure, W.R.: Mechanism of CRP-cAMP activation of *lac* operon transcription activation of the P1 promoter. J.Mol.Biol. **180**, 881-909, 1984.

(7) Busby, S., West., Lawes, M., Webster, C., Ishihama, A. & Kolb, A.: Transcritption activated by the *Escherichia coli* cyclic AMP receptor protein. Receptors bound in tandem at promoters can interact synergistically. J.Mol.Biol. **241**, 341-352, 1994.

(8)   Joung, J.K., Koepp, D.M. & Hochschild, A.: Synergistic activation of transcription by bacteriophage λ *CI* protein and *E. coli* cAMP receptor protein. Science **265**, 1863-1866, 1994.

(9)   Li, M., Moyle, H. & Susskind, M.M.: Target of transcriptional activation function of phage λ *CI* protein. Science **263**, 75-77, 1994.

(10) Kolkhof, P. & Müller-Hill, B.: λ *CI* repressor mutants altered in transcriptional activation. J.Mol.Biol. **242**, 23-36, 1994.

(11) Ryu, S., Garges, S. & Adhya, S.: An arcane role of DNA in transcription activation. Proc.Natl.Acad.Sci. USA **91**, 8582-8586, 1994.

(12) Tagami, H. & Aiba, H.: Role of CRP in transcription activation at *Escherichia coli lac* promoter: CRP is dispensable after the formation of open complex. Nucl.Acids Res. **23**, 599-605, 1995.

(13) Barberis, A., Pearlberg, J., Simkovich, N., Farrell, S., Reinagel, N., Bamdad, C., Sigal, G., & Ptashne, M.: Contact with a component of the polymerase II holoenzyme suffices for gene activation. Cell **81**, 359-368, 1995.

(14) Bracco, L., Kotlarz, D., Kolb, A., Diekmann, S. & Buc, H.: Synthetic curved DNA sequences can act as transcriptional activators in *Escherichia coli*. EMBO J. **8**, 4289-4296, 1989.

(15) Gartenberg, M.R. & Crothers, D.M.: Synthetic DNA bending sequences increase the rate of *in vitro* transcription initiation at the *Escherichia coli lac* promoter. J.Mol.Biol. **219**, 217-230, 1991.

(16) Busby, S. & Ebright, R.H.: Promoter structure, promoter recognition, and transcription activation in prokaryotes. Cell **79**, 743-746, 1994.

(17) Zinkel, S.S. & Crothers, D.M.: Catabolite activator protein-induced DNA bending in transcriptional initiation. J.Mol.Biol. **219**, 201-215, 1991.

(18) Kolb, A., Busby, S., Buc, H., Garges, S. & Adhya, S.: Transcriptional regulation by cAMP and its receptor protein. Ann.Rev.Biochem. **62**, 749-795, 1993.

(19) Crothers, D.M. & Steitz, T.A.: Transcriptional activation by *Escherichia coli* CAP protein. In: Transcriptional Regulation. Ed. by S.L. McKnight & K.R. Yamamoto. Cold Spring Habor Press, Cold Spring Harbor, N.Y., Vol. **1**, 501-534, 1992.

3.3.5 CAP in Other Systems

So far I have concentrated on CAP in the *lac* system. I have compared it to CAP in the *gal*, the *mel* and the *mal* systems. Like the *lac* operon, the *gal* operon is regulated by a repressor. This Gal repressor is homologous to Lac repressor (1). The *mel* operon looks like the *lac* operon. It consists of one gene coding for an α galactosidase and a second coding for α galactoside-permease. In analogy to the *lac* system, α galactosides are here hydrolysed but only the permease is structurally homologous to its *lac* counterpart. Both transport α- and β-galactosides (2). The *mel* system is positively regulated and its activator gene is weakly homologous to AraC, the positive regulator of the *ara* system. Finally, the *malT* gene codes for the specific activator of the *mal* (maltose) operons. At first glance one gets the impression that the function of CAP is essentially the same in all of these systems. But as the examination of these and other systems shows nothing could be farther from the truth.

Let me begin with the CAP gene itself. The CAP gene is autoregulated by CAP (3,4)! Like *lac* it has a site for activation -60.5 bp upstream of its transcription start. But in addition, it has a CAP site +42.5 bp downstream from the transcription start site which serves for repression! Repression guarantees that activation by $(CAP)_2cAMP$ will not be too strong (3,4).

Let us move to the second example, two of the eight operons controlled by CytR. Cyt repressor is homologous to Lac repressor (5). But it acts rather differently. At the *deo P2* promoter, CytR binds between two CAP proteins which are positioned -40.5 and -93.5 upstream of the transcription start (6). In the absence of CytR, the CAP bound at position -40.5 activates the *deo P2* promoter. In the presence of CytR, this activation is inhibited, but not by removing one or both of the CAP molecules. Instead a quaternary complex is formed consisting of DNA, the two CAP molecules with Cyt repressor located between them. This complex is unable to accomodate RNA polymerase at the adjacent *deo P2* promoter. If one destroys one of the CAP sites so that CAP cannot bind any longer, CytR does not bind well to "its" DNA (6). CAP mutants have been found which abolish Cyt repressor binding (7)! They occur in CAP codons 17 and 18, corresponding to the turn at the C-terminal end of helix A, and in codons 108 and 110, corresponding to the turn between helix B and C (see Fig. 10). Furthermore all constitutive CytR mutants occur in a region of CytR far away from its predicted DNA binding region.

The *deo* operon is one of eight operons controlled by Cyt repressor. The *cdd* operon is another one. Here, the situation is again different (8). CAP binds at positions -91.5 and -41.5 upstream of the *cdd* promoter. The CAP bound at position -91.5 helps to activate. But in the presence of *cytR*, CAP is removed to position -93.5. The resulting quaternary complex again inhibits the *cdd* promoter.

The third example is found in the *mal* system. The *mal* system essentially consists of two operons, the *malE* and the *malK* operon, which are transcribed from two divergent promoters positioned 271 bp apart. Four CAP sites are situated between the two promoters at positions -100.5, -132.5, -166.5 and -196.5 with respect to the *malK* promoter, or at positions -76.5, -105.5, -139.5 and -171.5 with respect to the *malE* promoter (9). All of these CAP sites are either too distant from each promoter, or in the case of -76.5 wrongly phased to be functional. Several MalT sites are positioned between the promoters and the CAP sites. In the absence of MalT activator CAP has no positive influence on these two promoters. MalT protein is also inactive in the absence of CAP. CAP and MalT together activate both promoters (9). What are the roles of CAP and MalT in this activation?

CAP seems to have two functions: first it repositions the MalT activator proteins (10). Then by bending the DNA it brings the MalT proteins into contact with RNA polymerase (11). Bending alone, without CAP, does not activate the promoters but together with MalT it does (11).

Finally, the case of the *pap* adhesin system (12) will be discussed. Again, we have two divergent promoters 331 bp apart. One CAP site is found between them at distances of -115.5 and -215.5 respectively. If this CAP site is destroyed, transcription of the two promoters is decreased more than tenfold. The presence of the product of the *papB* gene is also needed for effective transcription. In a mutant which lacks the nucleoid associated protein H-NS, the presence of both activators is not required. It seems that the two activators act together as antirepressors. Finally one may ask: the synthesis of how many proteins is activated or repressed by CAPcAMP? Protein production has been compared on two dimensional gels. About 20 proteins are decreased and another 30 increased by the presence of the CAP cAMP complex (13).

What is the bottom line? Proteins are versatile. Several of their domains may have been optimized for uses we do not understand, and they can be used for different purposes. But again, the number of optimal solutions is limited and can be analysed.

References

(1) Wilcken-Bergmann, B.V. & Müller-Hill, B.: Sequence of *galR* gene indicates a common evolutionary origin of *lac* and *gal* repressor in *Escherichia coli*. Proc.Natl.Acad.Sci. USA **79**. 2427-2431. 1982.

(2) Pourcher, T., Bassilana, M., Sarkar, H.K., Kaback, H.R. & Leblanc, G.: Melibiose permease and alpha-galactosidase of *Escherichia coli*: identification by selective labelling using a T7 RNA polymerase/promoter expression system. Biochemistry **29**, 690-696, 1990.

(3) Aiba, H.: Autoregulation of the *Escherichia coli crp* gene: CRP is a transcriptional repressor for its own gene. Cell **32**, 141-149, 1983.

(4) Hanamura, A. & Aiba, H.: A new aspect of transcriptional control of the *Escherichia coli crp* gene: positive autoregulation. Mol.Microbiol. **6**, 2489-2497, 1992.

(5) Valentin-Hansen, P., Larsen, L., Hojrup, J.E.P., Short, S. & Barbier, C.S.: Nucleotide sequence of the CytR regulatory gene of *E. coli* K12. Nucl.Acids Res. **14**, 2215-2228, 1986.

(6) Søgaard-Andersen, L., Pedersen, H., Holst, B. & Valentin-Hansen, P.: A novel function of the cAMP-CRP complex in *Escherichia coli*: cAMP-CRP functions as an adaptor for the CytR repressor in the *deo* operon. Mol.Microbiol. **5**, 969-974, 1991.

(7) Søgaard-Andersen, L., Mironov, A.S., Pedersen, H., Sukhodelets, V.V. & Valentin-Hansen, P.: Single amino acid substitution in the cAMP receptor protein specifically abolish regulation by the CytR repressor in *Escherichia coli*. Proc.Natl.Acad.Sci. USA **88**, 4921-4925, 1991.

(8) Holst, B., Søgaard-Andersen, L., Pedersen, H. & Valentin-Hansen, P.: The cAMP-CRP/CytR nucleoprotein complex in *Escherichia coli*: two pairs of closely linked binding sites for the cAMP-CRP activator complex are involved in combinatorial regulation of the *cdd* promoter. EMBO J. **11**, 3635-3643, 1992.

(9) Raibaud, O., Vidal-Ingigliardi, D. & Richet, E.: A complex nucleoprotein structure involved in activation of transcription of two divergent *Escherichia coli* promoters. J.Mol.Biol. **205**, 471-485, 1989.

(10) Richet, E., Vidal-Ingigliardi, D. & Raibaud, O.: A new mechanism for coactivation of transcription initiation: repositioning of an activator triggered by the binding of a second activator. Cell **66**, 1185-1195, 1991.

(11) Richet, E. & Søgaard-Andersen, L.: CRP induces the repositioning of MalT at the *Escherichia coli malKp* promoter primarily through DNA bending. EMBO J. **13**, 4558-4567, 1994.

(12) Forsman, K., Sondén, B., Göransson, M. & Uhlin, B.E.: Antirepression function in *Escherichia coli* for the cAMP-receptor protein transcriptional activator. Proc.Natl. Acad.Sci. USA **89**, 9880-9884, 1992.

(13) Mallick, U. & Herrlich, P.: Regulation of synthesis of a major outer membrane protein: cyclic AMP represses *Escherichia coli* protein III synthesis. Proc. Natl. Acad. Sci. USA **76**, 5520-5523, 1979.

3.4 Repression of the *lac* Promoter by Lac Repressor

3.4.1 The Modular Structure of Lac Repressor

The amino acid sequence of Lac repressor was determined in 1973 (1), the sequence of the *I* gene coding for it five years later (2). About ten grams of pure Lac repressor were used to determine its sequence, and about two hundred milligrams were given to various groups that tried to crystallize it for the determination of an X-ray structure. All this was in vain. Success in obtaining the first crystals of tetrameric Lac repressor complexed with *lac* operator DNA was announced only in 1990 (3). These crystals shatter when the inducer IPTG (isopropyl-1-thio-β-D-galactoside) is added to them (3). I recall here that the first explosion due to an allosteric change was reported by Felix Haurowitz: when deoxyhemoglobin crystals are exposed to oxygen, they shatter (4). Before the solution of the three crystal structures of Lac repressor, Lac repressor-IPTG complex and Lac repressor – *lac* operator complex was announced in 1996 (5), the crystal structure of Lac repressor core complexed with IPTG was published (6).

Lac repressor is a tetramer. Each subunit consists of three domains:

- Residues 1-59 constitute the headpiece which binds to DNA.
- Residues 60-330 constitute the core. The core binds IPTG and forms homodimers.
- Residues 331-360 are responsible for the aggregation of two active dimers to form tetramers.

The headpieces cannot be seen in the structure of Lac repressor and the Lac repressor-IPTG complex. They move. They can be seen only in the structure of the complex of Lac repressor with *lac* operator. An entire headpiece consists of four α helices connected by turns. The fourth helix lies in the minor groove between the major grooves occupied by the recognition helices. This structure very much resembles the structure of Pur repressor bound to its operator DNA in the presence of its corepressor (7). Pur repressor differs from Lac repressor in that it needs a corepressor to be brought into the proper conformation which binds specifically to *pur* operator DNA. The solution of the X-ray structure of unliganded Pur repressor which is unable to bind to *pur* operator allows one to understand the allosteric changes in the protein which determine the two states of Pur repressor (8).

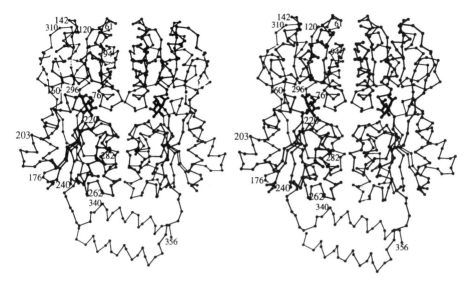

Fig. 12.: Tertiary structure of the complex of the core fragment of Lac repressor with inducer IPTG. Stereo drawing of the Cα backbone of one of the two dimers of the core
fragment. From Friedman *et al.*, 1995 (6), Fig. 2, with permission.

Pur repressor is one of the about ten known repressors which are structurally
related to Lac repressor (9). They each bind to different DNA targets. All of them
form only homodimers. That implies that they have different specificities in their
headpieces for DNA and different specificities in their cores for autodimerization.
Different small molecules stabilize their operator binding or their operator non-
binding form. Thus their inducer binding sites must differ too.

The first two α helices of the Lac repressor headpiece (and of all other structurally related repressors) form a helix-turn-helix (HTH) motif. The second helix
of the HTH motif is the recognition helix. In each dimer, the two recognition helices are 34 Ångström apart and fit into two consecutive major grooves in DNA.
In 1982 Brian Matthews, who had just solved the crystal structure of dimeric λ
Cro repressor, saw that Lac repressor has a HTH motif like Cro repressor, λ *CI*
repressor and other phage repressors. Therefore he proposed that a dimer of Lac
repressor recognizes operator DNA with the recognition helices of its HTH motives (10). Each subunit is thus capable of binding to one operator half.

In principle there are two ways in which a HTH motif can be arranged in a
repressor dimer:

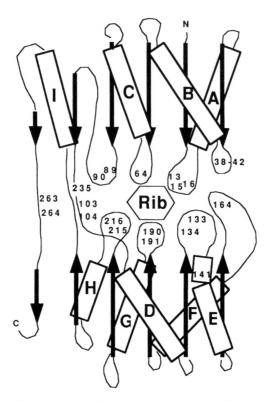

Fig. 13.: Cartoon showing the location of sugar binding residues in the ribose receptor of
E.coli with respect to structural topology. From (18), Fig. 9, with permission.

- The N-terminus of the helix preceding the recognition helix may point away
  from the center of symmetry of the operator. Such is the case in λ *cro* and the
  other phage repressors.
- The N-terminus of the helix preceding the recognition helix may point towards
  the center of symmetry of the operator. Such is the case with Lac repressor. This
  was demonstrated by Robert Kaptein's NMR analysis of the complex of Lac
  repressor headpiece and a half operator (11).

This structure has been refined and refined. I will only quote the latest paper (12).
The short headpiece was first used. It ends with residue 53. So the HTH motif
and a third helix could be seen. The fourth helix, which begins at residue 50 and
ends at residue 58, cannot be seen in the NMR analysis. According to the X-ray

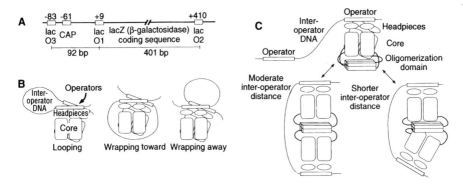

Fig. 14.: Schematic diagrams of higher order interactions between Lac repressor and the *lac* operon. (A) The control sites of the *lac* promoter. (B) Possible modes of bidentate interaction between two *lac* operators and Lac repressor. (C) Postulated partial unfolding of Lac repressor and subsequent segmental flexibility in binding two *lac* operators separated by distances shorter than can be accomodated by the repressor in its crystal conformation without severe strain. From (6), Fig.7, with permission.

structure (5) the fourth helices of two subunits pass through the central minor groove of *lac* operator.

When the sequence of the core of Lac repressor was compared to the sequences of three different sugar (galactose, arabinose, ribose) binding, periplasmic proteins, weak homology was discovered (13). The sugar binding proteins are involved in sugar transport and in chemotaxis as chemoreceptors. The crystal structure of one of them had already been solved when the sequence comparison was made (14). In subsequent years the crystal structures of several more periplasmic proteins involved in chemotaxis were solved and all were found to be similar (15-16). This is astonishing since two of them, the leucine (16) and lysine (17) receptors, show no homology in their protein sequences with the sugar receptors. All receptors have similar three-dimensional structures and these structures are similar to the structures of the cores of Lac (6) and Pur (7,8) repressors!

Their structures consist essentially of nine interconnected β-sheet-α-helix (β-α) elements. These elements form two similar domains. Between the domains the effector molecule is bound (Fig. 13). Two subunits are needed to bind specifically to an operator with dyadic symmetry. Binding is substantially stronger if the two subunits form a suitably aligned dimer. Lac and Pur repressor use side chains of three α helices in the C-terminal domain of the core to form specific homodimers.

Lac repressor is, in contrast to all other related repressors, a stable tetramer. It differs from all other repressors of its family in its structure by two properly presented leucine heptad repeats at its extreme C-terminus, residues 331-360. As predicted (21) they form an antiparallel, four helical bundle (5,6). But in contrast to earlier expectations (22) the two dimers are arranged in a V form, head to head and nearly parallel to each other (Fig. 14).

References

(1) Beyreuther, K., Adler, K., Geisler, N. & Klemm, A.: The amino-acid sequence of *lac* repressor. Proc.Natl.Acad.Sci. USA **70**, 3576-3580, 1973.

(2) Farabaugh, K.: Sequence of the *lacI* gene. Nature **274**, 765-769, 1978.

(3) Pace, H.C., Lu, P. & Lewis, M.: *lac* repressor: Crystallization of intact tetramer and its complexes with inducer and operator DNA. Proc.Natl.Acad.Sci. USA **87**, 1870-1873, 1990.

(4) Haurowitz, F.: Das Gleichgewicht zwischen Hämoglobin und Sauerstoff. Hoppe-Seyler's Z.Physiol.Chem. **254**, 268-274, 1938.

Max Perutz recalls, in the introduction to his book "Mechanisms of cooperativity and allosteric regulation in proteins" (Cambridge University Press, Cambridge, 1990): "Nearly fifty years later I was in the United States and telephoned Haurowitz at his home in Bloomington, Indiana, to ask if he would like me to visit him. He told me to come, because he had something very important to tell me. He was then over ninety and seriously ill, which made me wonder if he had some last message to give me about his theories of immunology. In fact, he talked about that paper. He told me that when he sent it to Hoppe-Seyler's *Zeitschrift*, German journals were forbidden to publish papers by Jewish authors like himself, but the editors, F. Knoop in Freiburg and K. Thomas in Leipzig, ignored that prohibition and published his paper regardless of the Nazi's racial laws. Haurowitz said that Knoop was actually imprisoned for his offense. Haurowitz died a few months after my visit. This preface gives me the opportunity of fulfilling his last wish that I should pay tribute to Knoop's and Thomas's courageous stand in the face of Nazi terror, which secured the publication of the first experimental evidence of an allosteric change in protein." Knoop and Thomas were undoubtedly decent and courageous. But according to documents in the Berlin Document Center and the Tübingen University Archives the truth was never actually divulged so Knoop was never reprimanded or imprisoned.

(5) Lewis, M., Chang, G., Horton, N.C., Kercher, M.A., Pace, H.C., Schumacher, M.A., Brennan, R.G. & Lu, P.: Crystal structure of the *Escherichia coli* lactose operon repressor and its complexes with DNA and inducer. Science **271**, 1247-1254, 1996.

 (6) Friedman, A.M., Fischmann, T.O. & Steitz, T.A.: Crystal structure of *lac* repressor core tetramer and its implications for DNA looping. Science **268**, 1721-1727, 1995.

 (7) Schumacher, M.A., Choi, K.Y., Zalkin, H. & Brennan, R.G.: Crystal structure of LacI member, PurR, bound to DNA: minor groove binding by α helices. Science **266**, 763-770, 1994.

 (8) Schumacher, M.A., Choi, K.Y., Lu, F., Zalkin, H. & Brennan, R.G.: Mechanism of corepressor-mediated specific DNA binding by the purine repressor. Cell **83**, 147-155, 1995.

 (9) Markiewics, P., Kleina, L.G., Cruz, C., Ehret, S. & Miller, J.H.: Genetic studies of the *lac* repressor, XIV. Analysis of 4.000 altered *Escherichia coli lac* repressors reveals essential and nonessential residues, as well as "spacers" which do not require a specific sequence. J.Mol.Biol. **240**, 421-433, 1994.

(10) Matthews, B.W., Ohlendorf, D.H., Anderson, W.F. & Takeda, Y.: Structure of the DNA-binding region of *lac* repressor inferred from its homology with *cro* repressor. Proc.Natl.Acad.Sci. USA **79**, 1428-1432, 1982.

(11) Boelens, R., Scheek, R.M., van Boom, J.H. & Kaptein, R.: Complex of *lac* repressor headpiece with a 14 base-pair *lac* operator fragment studied by two-dimensional nuclear magnetic resonance. J.Mol.Biol. **193**, 213-216, 1987.

(12) Slijper, M., Bonvin, A.M.J.J., Boelens, R. & Kaptein, R.: Refined structure of *lac* repressor headpiece (1-56) determined by realaxation matrix calculations from 2D and 3D NOE data: change of tertiary structure upon binding to the *lac* operator. J.Mol.Biol. **259**, 761-773, 1996.

(13) Müller-Hill, B.: Sequence homology between Lac and Gal repressor and three sugar-binding periplasmic proteins. Nature **302**, 163-164, 1983.

(14) Gilliland, G.L. & Quiocho, F.A.: Structure of the L-arabinose-binding protein from *Escherichia coli* at 2.4Å resolution. J.Mol.Biol. **146**, 341-362, 1981.

(15) Vyas, N.K., Vyas, M. & Quiocho, F.A: Sugar and signal-transducer binding sites of the *Escherichia coli* chemoreceptor protein. Science **242**, 1290-1295, 1988.

(16) Sack, J.S., Trakhanov, S.D., Tsigannik, I.H. & Quiocho, F.A.: Structure of L-leucine-binding protein refined at 2.4Å resolution and comparison with the leu/ile/val-binding protein structure. J.Mol.Biol. **206**, 193-207, 1989.

(17) Kang, C.H., Shin, W.C., Yamagata, Y., Gokcen, S., Ames, G.F.L. & Kim, S.H.: Crystal structure of the lysine-, arginine-, ornithine-binding protein (LAO) from *Salmonella typhimurium* at 2.7Å resolution. J.Biol.Chem. **266**, 23893-23899, 1992.

(18) Mowbray, S.L. & Cole, L.B.: 1.7Å X-ray structure of the periplasmic ribose receptor from *Escherichia coli*. J.Mol.Biol. **225**, 155-175, 1992.

(19) Sams, C.F., Vyas, N.K., Quiocho, F.A. & Matthews, K.S.: Predicted structure of the sugar-binding site of *lac* repressor. Nature **310**, 429-430, 1984.

(20) Nichols, J.C., Vyas, N.K., Quiocho, F.A. & Matthews, K.S.: Model of lactose repressor core based on alignement with sugar-binding proteins is concordant with genetic and chemical data. J.Biol.Chem. **268**, 17602-17612, 1993.

(21) Alberti, S., Oehler, S., Wilcken-Bergmann, B.v. & Müller-Hill, B.: Genetic analysis of
the leucine heptad repeats of Lac repressor: evidence for a 4-helical bundle. EMBO J. **12**,
3227-3236, 1993.
(22) McKay, D.B., Pickover, C.A. & Steitz, T.A.: *Escherichia coli lac* repressor is elongated
with its operator DNA binding domains located at both ends. J.Mol.Biol. **156**, 175-183,
1982.

3.4.2 Mutational Analysis of the *lac* Operators *O1*, *O2* and *O3*

The Jacob-Monod model of repression and the evidence for it were straightforward (1). Lac repressor binds to *lac* operator and thereby inhibits transcription. Mutants in which Lac repressor is destroyed are recessive. The presence of Lac repressor is dominant over its absence. Mutants in which *lac* operator is destroyed are *cis*-dominant, *i.e.* they are dominant to the operon of which the operator is destroyed. Initially and also later O^c mutants were *only* found in one location positioned around its symmetry center nine bp downstream of the transcription start site of the *lac* promoter. Steric hindrance of RNA polymerase by Lac repressor was a most plausible mechanism of repression (2).

In 1974 two DNA sequences were identified which specifically bound Lac repressor and which had some sequence homology with the main operator (*O1*). One sequence (*O3*) mapped 82 bp upstream (2), the other sequence (*O2*) mapped 410 bp downstream of the transcription start site (2,3). Their function was unclear. It was argued that if they had some vital function, mutants should have been found which mapped there and which led to a constitutive phenotype. But all screens for operator mutants in *O3* or *O2* were in vain. Apparently such mutants did not exist. Moreover, the two *pseudooperators*, as they were called, differed from *O1* in quality. *O2* has about ten percent of the binding capacity of *O1* and *O3* has much less. Today we estimate that *O3* has only about 0.3 percent binding capacity of *O1*. It is a very weak *lac* operator.

When *O2* and *O3* were identified, it was not clear how Lac repressor binds to *lac* operator. It took some time until it was shown (but not generally accepted, see section 2.9.) that two subunits of one Lac repressor tetramer bind to one *lac* operator (4,5). The first evidence for DNA-protein-DNA loops was found in the *ara* system by Robert Schleif (6,7). Later, Tom Record placed a wildtype *lac* operator at 118, 185 or 283 bp upstream of a 40 bp *lac* promoter fragment carrying a *lac* O^c operator (8). In these constructs *O3* and *O2* were absent. The *lacZ* gene was replaced by DNA coding for the easily testable galactokinase. At the distance of

118 bp a twofold increase in repression was observed in the plasmid constructs. The two other constructs led to a tenfold increase in repression.

Only in 1990 was it realized that it makes no sense to analyse one auxiliary operator in the presence of the other. Moreover, the analysis must be done with chromosomal constructs, since plasmid constructs show only weak effects. The relevant experiments were performed by Stefan Oehler (9,10). He destroyed both *O3* and *O2* on a chromosomal construct. In so doing he discovered that *O3* and *O2* are *redundant*, to use the pejorative term often used in mouse knock-out genetics. Both *O3* and *O2* are capable of forming a loop with tetrameric Lac repressor and *O1*. Either a loop *O3*-LacR-*O1* or a loop *O2*-LacR-*O1* is present. Each loop has about the same effect. Each increases repression about thirty fold. Where both *O2* and *O3* are present, repression of *O1* increases to about seventy fold above the seventeen fold repression found in their absence. Thus, in the absence of both *O2* and *O3*, repression of ordinary *O1* decreases to less than two percent of the normal wildtype level! But if only one of the auxiliary operators is destroyed, a marginal threefold effect is observed, a difference that genetic screens would never detect! Thus *O2* and *O3* are truly *auxiliary* operators and not, as they were called by their discoverers, *pseudo* (lying, false) fake-operators.

How do the three operators *O1*, *O2* and *O3* act? Operator *O1* acts essentially as originally proposed (2). Lac repressor bound to *O1* competes with RNA polymerase for effective binding. The more Lac repressor bound, the stronger the repression (11). The less Lac repressor bound, the weaker the repression. Thus, repression depends on the concentration of Lac repressor and the quality of *O1*. Repression also depends on the quality of the promoter. The better the promoter, the weaker the repression (12) (see also section 2.15.).

The auxiliary operators serve to increase the local concentration of Lac repressor at *O1*. In the wildtype situation, about ten molecules of Lac repressor tetramer are present in each cell. This corresponds to a repressor concentration of about 10^{-8}M (in fact at least 90 percent of Lac repressor is bound nonspecifically to DNA (13); this does not change the argument). If *O2* is occupied by tetrameric Lac repressor, *O2* binds to only one of the dimers of the repressor tetramer (4,5); the other dimer is still available for DNA-binding. If we assume that this unoccupied dimer moves freely within a sphere with a radius of the distance between *O1* and *O2* (the actual radius will be smaller), then its concentration is increased by about 20 fold over the normal concentration of Lac repressor (10^{-8}M), as measured at *O1*. This should lead to 20 fold tighter binding of this repressor dimer to *O1* through operator-repressor-operator loop formation. Repression should thus increase by at least 20 fold.

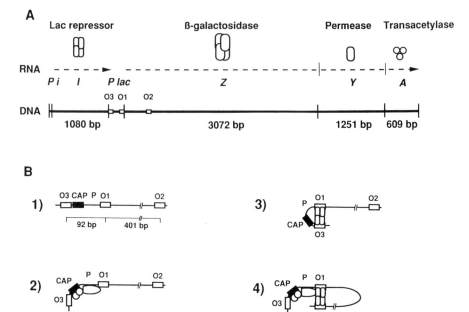

Fig. 15.: Model of Lac repression. A. The *lac* system. B. 1. The *lac* promoter region. 2. Un-
          repressed transcription. 3. Repression by tetrameric Lac repressor binding to *O1 and*
          O3. The CAP site and the *P1* promoter are depicted as being unoccupied. There is
          no evidence for or against this. 4. Repression by tetrameric Lac repressor binding to
          O1 and O2. It is likely but not explicitly demonstrated that CAP protein and RNA
          polymerase are bound to their sites. From (10), Fig. 1, with permission.

I have assumed here that *O2* is one hundred percent occupied. In fact the binding constant of *O2* to Lac repressor has been estimated to be about ten fold lower
than the binding constant of *O1* (9,10). Therefore *O2* is occupied only to about
50 percent. This decreases the expected increase in local concentration, and thus
repression, two fold. In fact a thirty fold decrease of repression is what one observes if one destroyes *O2* in the absence of *O3* (9,10). This should remind us
that auxiliary operators need not bind to repressor with 99.9 percent efficiency.
For proper functioning of *O2*, a 50 percent binding efficiency is clearly sufficient.

If we measure the effect of the destruction of *O2* upon repression in the wild
type system, *i.e.* in the presence of *O3*, only about a three fold difference is seen
(9,14). Why is this so? Because the other auxiliary operator, *O3*, complements
the action of *O2*. Note *O3* is positioned 82 base pairs upstream of the transcrip-

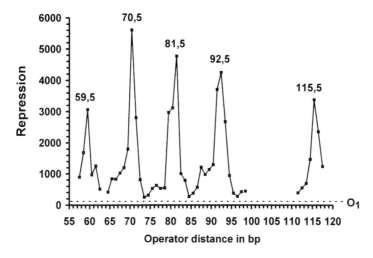

Fig. 16.:   Repression depends on the exact distance between *O1* and the auxiliary operator. A
synthetic *lac* UV5 system is used, which lacks the CAP and the *O2* sites. An ideal
operator is used as auxiliary operator, *O1* as main operator. From (16), with permission.

tion start. If we calculate the increase in local Lac repressor concentration at *O1*
within a sphere with a radius of the distance of *O1* to *O3*, we find a huge increase
of 2×10^4. This huge factor is decreased by two counteracting factors: 1. by the
energy which is needed to bend the DNA; 2. by the very low binding constant of
O3 for Lac repressor. The binding is only one part in 300 compared to *O1*. We
calculate from this that only about 5 percent of *O3* are normally occupied by Lac
repressor *in vivo*. Thus repression would increase about twenty fold if *O3* were
optimised in its sequence. Indeed this is what one observes (10).

So far I have disregarded phasing. Whatever the exact form of the loop (see
Fig. 14) there must be optimal distances between the two operators such that the
loop can be formed with minimal strain (15,16). I show here the measurements
made by Johannes Müller (16) (Fig. 16). The optima become more and more pronounced the closer the two operators are placed. This also indicates that the operators are at (almost) *optimal* positions in the *lac* O1-O3 and in the *gal* system.
The distance of 92.5 bp between *O1-O3* is as optimal as the distance between
the two *gal* operators, *i.e.* 114 bp. This may suggest planning by Dawkin's blind
watchmaker. Rather it is selection for the optimal place.

In summary *both O2* and *O3* have to be destroyed so that the effect of their action can be seen. When this is done we note an about seventy fold decrease in repression compared to the wild type situation. In the jargon now used in mouse genetics, *O2* and *O3* are *redundant*. Using classical genetic methods one has never obtained nor will ever obtain mutants with constitutive phenotypes in them. Any mutant in one operator is rescued by the existence of the other operator. The word *redundant* better characterizes present day mentality than the facts. In this context it will no longer be used. I will also not use the name pseudooperators, lying, false operators given to *O2* and *O3* by their discoverers, who suspected that they had no function. This is why I prefer to call them auxiliary (helping) operators (9,10).

Finally the question may be asked: has anybody seen operator-repressor-operator loops? Indeed such loops have been seen in EM pictures (17), evidence was found for such loops in gel experiments (17,18) and finally the kinetics of the formation and breakdown of loops formed in single molecules has been measured (19).

References

(1) Jacob, F. & Monod, J.: Genetic regulatory mechanisms in the synthesis of proteins. J.Mol.Biol. **3**, 318-356, 1961.

(2) Gilbert, W., Gralla, J., Majors, J. & Maxam, A.: Lactose operator sequences and the action of *lac* repressor. In: Symposium on Protein-Ligand Interactions. Ed. by H. Sund & G. Blauer. Walter de Gryuter, Berlin, 193-206, 1975.

(3) Reznikoff, W.S., Winter, R.B. & Hurley, C.K.: The location of the repressor binding sites in the *lac* operon. Proc.Natl.Acad.Sci. USA **71**, 2314-2318, 1974.

(4) Kania, J. & Brown, D.T.: The functional repressor parts of a tetramer *lac* repressor-β-galactosidase chimaera are organized as dimers. Proc.Natl.Acad.Sci. USA **73**, 3529-3533, 1976.

(5) Kania, J. & Müller-Hill, B.: Construction, isolation and implications of repressor-galactosidase-β-galactosidase hybrid molecules. Eur.J.Biochem. **79**, 381-386, 1977.

(6) Dunn, T.M., Hahn, S., Ogden, S. & Schleif, R.F.: An operator at -280 base pairs that is required for repression of *araBAD* operon promoter: addition of DNA helical turns between the operator and promoter cyclically hinders repression. Proc.Natl.Acad.Sci. USA **81**, 5017-5020, 1984.

(7) Schleif, R.F.: DNA looping. Science **240**, 127-128, 1988.

(8) Mossing, M.C. & Record, M.T.: Upstream operators enhance repression of the *lac* promoter. Science **233**, 889-892, 1986.

(9) Oehler, S., Eismann, E.R., Krämer, H. & Müller-Hill, B.: The three operators of the *lac* operon cooperate in repression. EMBO J. **9**, 973-979, 1990.

(10) Oehler, S., Amouyal, M., Kolkhof, P., Wilcken-Bergmann, B.v. & Müller-Hill, B.: Quality and position of the three *lac* operators of *E. coli* define the efficiency of repression. EMBO J. **13**, 3348-3355, 1994.

(11) Schlax, P.J., Capp, M.W. & Record, M.T.: Inhibition of transcription initiation by *lac* repressor. J.Mol.Biol. **245**, 331-350, 1995.

(12) Lanzer, M. & Bujard, H.: Promoters largely determine the efficiency of repressor action. Proc.Natl.Acad.Sci. USA **85**, 8973-8977, 1988.

(13) Kao-Huang, Y., Revzin, A., Butler, A.P., O'Conner, P., Noble, D.W. & von Hippel, P.H.: Nonspecific DNA binding of genome regulating proteins as a biological control mechanism: measurement of DNA-bound *Escherichia coli lac* repressor *in vivo*. Proc.Natl.Acad.Sci. USA **74**, 4228-4232, 1977.

(14) Eismann, E., Wilcken-Bergmann, B.v. & Müller-Hill, B.: Specific destruction of the second *lac* operator decreases repression of the *lac* operon in *Escherichia coli* five fold. J.Mol.Biol. **195**, 949-952, 1987. The reader of this paper will discover that the text and its data say that repression was decreased by three to five fold. At the time we, like all others, did not appreciate the function of *O3*. I pushed the five fold of the title. I am embarrassed.

(15) Law, S.M., Bellomy, G.R., Schlax, P.J. & Record, M.T.Jr.: *in vivo* thermodynamic analysis of repression with and without looping in *lac* constructs. J.Mol.Biol. **230**, 161-173, 1993.

(16) Müller, J., Oehler, S. & Müller-Hill, B.: Repression of *lac* promoter as a function of distance, phase and quality of an auxiliary *lac* operator. J.Mol.Biol. **257**, 21-29, 1996.

(17) Krämer, H., Niemöller, M., Amouyal, M., Revêt, M., Wilcken-Bergmann, B.v. & Müller-Hill, B.: *lac* repressor forms loops with linear DNA carrying two suitably spaced *lac* operators. EMBO J. **6**, 1481-1491, 1987.

(18) Krämer, H., Amouyal, M., Nordheim, A. & Müller-Hill, B.: DNA supercoiling changes the spacing requirement of two *lac* operators for DNA loop formation with *lac* repressor. EMBO J. **7**, 547-556, 1988.

(19) Finzi, L. & Gelles, J.: Measurement of repressor-mediated loop formation and breakdown in single DNA molecules. Science **267**, 378-380, 1995.

3.4.3 Mutational Analysis of Lac Repressor

3.4.3.1 General Mutant Analysis: Nonsense and Missense Mutants of the *lacI* Gene

Lac repressor was and still is the best genetically analysed protein. Jeffrey Miller replaced first 141 codons (1), and then every codon (2) between codon 2 and 329 of the 360 codons of the *lacI* gene with an amber codon. He then tested every amber mutant with thirteen natural or synthetic amber suppressors, and determined whether the suppressed amber mutant was inducible (I^+), constitutive (I^-) or non-inducible (I^s).

The synthetic amber supressors had been prepared by Jeffrey Miller and others (3-5). They replaced the natural anticodon in various transfer RNA genes with the amber anticodon. One might think that one should be able to do this successfully with *every* transfer RNA gene. Unfortunately this is not the case. In seven cases the tRNA synthetases specifically recognizes the anticodon and so one cannot touch it easily without destroying the loading specificity of these tRNAs. So the missing seven amino acids can not be inserted by this method. Unfortunately Miller's paper (2) does not contain a table with all the suppression data. This list will appear shortly elsewhere (6).

The reader may recall that Jeffrey Miller had mapped 2,000 constitutive and noninducible missense mutants (7). He never got around to sequencing them. Barry Glickman did so. He and his collaborators sequenced the DNA of over 5,000 *lacI⁻* constitutive mutants (8). They uncovered 189 missense mutations generating 184 amino acid substitutions at 103 different sites within the Lac repressor. Miller and Glickman were geneticists who have spent a considerable part of their own and their student lives in amassing these data. Have these data ever been used by protein chemists to make predictions within a theory of protein structure? The answer is NO. The data wait to be discovered for intelligent utilisation.

3.4.3.2 Special Mutant Analysis: Specific Protein-DNA Recognition

In 1982 Brian Matthews predicted that Lac repressor has a helix-turn-helix (HTH) motif and that its recognition helix begins with residue 17 (9). Three years later his prediction was verified by NMR analysis of the Lac repressor headpiece (10). About this time the techniques of reverse genetics became available. So

Richard Ebright (11, 12) first replaced Gln 18 (the second residue of the recognition helix) of Lac repressor with glycine, serine or leucine. He then constructed six *lac* operator mutants which had single substitutions in base pairs 1 to 6 of *lac* operator (I use the systematic numbering system for O^c mutants generally used and not the specific one of Ebright).

```
        7 6 5 4 3 2 1
5'      T G T G A G C G G A T A A C A      3'
3'      A C A C T C G C C T A T T G T      5'


        A C T G T T
        T G A C A A
```

Wildtype Lac repressor represses wildtype *lac* operator by a factor of about one thousand. It represses the various operator mutants 20 to 100 times less. The three mutant Lac repressors repress wildtype *lac* operator about 20 to 100 times less than wild type repressor. What happens if one uses a mutant Lac repressor to repress the mutant *lac* operators? The two effects should multiply, unless the particular base pair which has been substituted in the O^c mutant interacts with the amino acid of Lac repressor which has also been replaced. Then, and only then, the effects should not multiply, since it does not matter that *both* partners have been substituted. In the case of Ebright's experiment the outcome was unambiguous. Base pair 5 (my numbering!) was predicted to interact with residue 2 of the recognition helix. Thus Ebright predicted a specific amino acid – base pair interaction from his genetic data. This prediction was confirmed by the NMR structure of Robert Kaptein two years later (13).

There is another genetic way of exploring protein-DNA recognition. This is to look for specificity changes. If a particular O^c mutant is not recognized (*i.e.* repressed) well by wild type Lac repressor, one may find a particular mutant Lac repressor which binds (represses) this particular O^c mutant well but which does not bind (repress) well to wild type operator. How does one go about finding these rare cases? There are two ways of solving this problem. 1. One tries model building and makes an educated guess. 2. One examines homologous repressors and compares their recognition helices and operators with those of the *lac* system. Since they have different functions, very likely they have different specificities. For example Gal repressor has the sequence Val, Ala in positions 1 and 2 of its recognition helix, in contrast to Lac repressor which at these positions has the sequence Tyr Gln. The main difference between the two operators is at position 4:

```
    -7-6-5-4-3-2-1   1 2 3 4 5 6 7
5'   T G T G A G C - G C T C A C T 3'   ideal lac operator
5'   T G T A A X C - G X T T A C T 3'   consensus gal operator
```

The corresponding variants of Lac repressor and *lac* operator were tested (14). A strong specificity change is indeed observed. This proves that residues 1 and 2 of the recognition helix are involved in the recognition of base pair 4. This type of analysis can and has been extended to other systems, including the λ system (15).

Lac and λ *CI* or λ cro repressor indeed have a surprising similarity. Their recognition helices recognize the sequence 5' GTGA 3', but their subunits are assembled in an opposite manner (15, 16). This can be best seen in Fig. 17. Here one sees that the right subunit of Lac repressor equals the left subunit of λ cro or λ *CI* and vice versa. To remember this fact, it is best to recall that the helix preceding the recognition helix of Lac repressor points towards the center of symmetry of *lac* operator. In λ *CI* repressor or λ *cro* it points away from the center of symmetry of λ operator.

A comparison of the repressor complexes binding with HTH motifs to their respective operators indicates that the interactions with the phosphates and bases are not identical. They are only similar (Fig. 18). Yet the similarity suffices for the HTH motif of a tight binding variant of Lac repressor to be replaced successfully with the HTH motif of λ *cro* (16).

To test substitutions according to existing homologous systems is fruitful, but one is limited by the number of described homologous systems. Ideally one would like to know the entire potential of the four residues of the recognition helix responsible for base pair recognition *i.e.* the potential of residues 1, 2, 5 and 6 of the recognition helix. All amino acids in positions 1, 2 and 6 were tested (17, 18). Their effects were additive. So, Lac repressors were constructed which recognize previously unknown operators. The results for residues 1 and 2 and base pairs 4 and 5 are summarized in Fig. 19. Here we see that only some amino acids are used to recognize base pairs. Not all base pairs can be effectively recognized by the existing amino acids.

The analysis of residues which are additive in their function is simple. This is indeed the case for residues 1, 2 and 6 of the recognition helix of Lac repressor. In contrast residue 5 interacts with residues 1 and 2. This destroys the principle of additivity and yields unpredicted results (19).

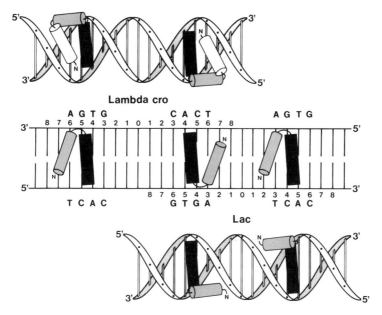

Fig. 17.: Schematic representation of the operator complexes of λ cro protein (or λ *CI* repressor) and Lac repressor. The two dimensional model illustrates how the dimerization interfaces of the repressors determine the orientation of the two HTH motifs toward each other. Note that the upper strand is written from 3' to 5' to allow the recognition helices to lie in front of the double helix. The recognition helices are depicted in black, the helices preceding the recognition helices in grey. The *lac* operator shown here carries a central base pair. Only those base pairs which presumably are recognized by the side chains of the recognition helices are indicated. From (16), with permission.

3.4.3.3 Special Mutant Analysis: Specific Protein-Inducer Interaction

A comparison of the I^s mutants generated by Miller (1, 2) with the X-ray structures (see section 3.4.1. and 3.4.3.) of Lac repressor indicates that every single residue invoked in inducer recognition can be turned into an I^s mutant (6). This is expected. The challenge of replacing these and other residues in order to construct binding sites for new, radically different inducers, has not been met.

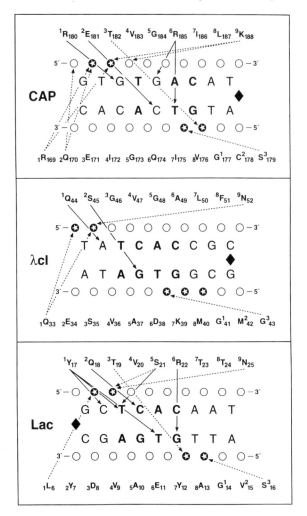

Fig. 18.: Schematic diagrams of the interactions of side chains of the HTH motives of CAP protein, λ *CI* repressor and Lac repressor with their DNA targets (operators). In the case of the CAP target and λ operator their left half sites are shown, in the case of *lac* operator its right half site. The residues on top are the residues of the recognition helix. The residues below are the residues of the preceeding helix and the turn. The numbers next to the residues indicate: the number in the intact protein (below, right); the number in the recognition helix (on top, left); the number is the preceding helix (below, left); and the number in the turn (on top, right). The interactions have been deduced from X-ray, NMR, and genetic-biochemical data.

lac ideal Operator

8 7 6 5 4 3 2 1   1 2 3 4 5 6 7 8
5′ T T G|T G|A G C – G C T C A C A A 3′

Recognition helix of wt lac Repressor

1 2 3 4 5 6 7 8 9
|Y Q|T V S R V V N

4＼5	A	C	G	T	residue
A	I, S		A, S	P, I, S	1
	S, A, C, I, T, V		T	A, G, V	2
C	P, M				1
	K				2
G			K	S, I, H, Y, R	1
			S	M, Q	2
T	P, S, V		K, P, S, T, V	H, S	1
	M		M, S, T	M	2

Fig. 19.: The Lac repressor and its mutants in residues 1 and 2 of the recognition helix which repress at least 150 fold the ideal *lac* operator and its variants in base pairs 4 and 5. Upper line: residue 1. Lower line: residue 2. Any combination of these two residues represses the particular operator at least 150 fold in the test system used. For seven of the *lac* operator variants no such combination exists. From (18), with permission.

3.4.3.4 Special Mutant Analysis: Altered Allosteric Properties

Two types of mutants can be predicted: 1. mutants which are fixed in the operator-binding conformation. They cannot be induced. Thus, they have the phenotype of I^s mutants. Miller's mutants (1, 2) have been analysed and have been listed (6). Yet a mechanistic understanding of these mutants is lacking. 2. mutants which are fixed in the inducer-binding conformation. They are constitutive. They should bind inducer *in vitro*. However again a detailed analysis is missing.

Here I mention only one mutant which has been analysed in detail (20): the I^r mutant used to isolate Lac repressor (see section 1.13.). It binds the inducer IPTG both *in vivo* and *in vitro* about two times better than wild type. It carries an Ala -> Thr substitution at residue 110. It has been shown that in this mutant Lac repressor the equilibrium between the operator and inducer binding form has been shifted toward the inducer-binding form. An Ala -> Lys substitution at the same position leads to freezing of the operator binding conformation. Model building shows this convincingly.

3.4.3.5 Special Mutant Analysis: Altered Dimerization and Tetramerization

Lac repressor is a tetramer in the concentration at which it occurs in the *E. coli* cell. The dyad symmetry of the *lac* operator demands that two monomers form a dimer, which then bind to operator. This interaction must be specific and tight. Otherwise mixed dimers among the various members of the Lac repressor family (GalR, PurR, RafR etc.) would form and mess up all specific regulation. The monomer-monomer interaction occurs in that part of the interface of the core of Lac repressor which is removed from the headpiece (see Fig. 12). This fits Jeffrey Miller's prediction that residues near positions 220 and 280 are involved in dimer formation (21). So far nothing has been published on attempts to change the specificity of this monomer-monomer interaction. Unpublished work hints that suitable single substitutions of residues in these regions change the specificity of dimerization (22).

The selection systems which in general used a tenfold increase in amounts of Lac repressor (I^Q) are incapable of allowing the isolation of a mutant which destroys the capability of Lac repressor dimers to form tetramers. The dimeric Lac repressor represses well enough under these conditions. So, how was the first mutant unable to form tetramers discovered? This is an interesting story which merits reporting.

Stefan Oehler, an undergraduate student working in the Cologne Genetics Institute, was asked as his first experiment to do gel shifts with Lac repressor. He was given a DNA fragment which carried two suitable *lac* operators separated by a suitable distance, and a plasmid which overproduced Lac repressor. As has been discussed (section 3.4.2.), with such DNA Lac repressor forms loops. In acrylamide gels these loop-complexes move slower than all other possible complexes. So, Oehler was asked to produce loops on gels. But no matter what he did, he observed only complexes which moved to the position where one expects single or double occupancy. He observed not the slightest sign of loop formation. At first, everybody thought this is the typical accident of a beginner. But when he reproduced his gels the second and third time, doubts were raised as to whether Oehler was right in his mind! It took some time until it was realized and proven that the particular *lacI* gene which Oehler used for his experiments had acquired a mutation during cloning which led to this phenotype. Oehler showed that it was a frameshift in codon 330 (15). The DNA near codon 330 had not been manipulated. The mutation was a present from heaven. It was received gratefully, but it also gave the uneasy feeling that such mutants do accumulate all the time everywhere in the DNA with which one works, and that most of the time their effects are destructive and misleading.

The plasmid had been used before in the lab and had been sent to several groups as a good source of tetrameric wild type Lac repressor. Immediately those people who had received it were informed by letter and by a note in a publication (15). And, of course, if one inspected the sequence between codon 330 and 360 it was evident that two leucine repeats were present. So, two US groups immediately began to analyse the system (23-25). I am not quoting their work in detail here, since the analysis done in Cologne was, after all, more informative. In a first paper all relevant residues of the two heptad repeats were replaced and the effect upon dimerization was measured (26). In a second approach, experiments were done which suggested that the four C-terminal alpha-helices of Lac repressor assemble as a four helical bundle (27). Two years later this was verified by the X-ray structure of the core (Fig. 14).

References

(1) Kleina, L.G. & Miller, J.H.: Genetic studies of the *lac* repressor. XIII. Extensive amino acid replacements generated by the use of natural and synthetic nonsense suppressors. J.Mol.Biol. **212**, 295-318, 1990.

(2) Markiewicz, P., Kleina, L.G., Cruz, C., Ehret, S. & Miller, J.H.: Genetic studies of the *lac* repressor. XIV. Analysis of 4000 altered *Escherichia coli lac* repressors reveals essential and non-essential residues, as well as "spacers" which do not require a specific sequence. J.Mol.Biol. **240**, 420-433, 1994.

(3) Normanly, J., Masson, J.M., Kleina, L.G., Abelson, J. & Miller, J.H.: Construction of two *Escherichia coli* amber suppressor genes: tRNA Phe/Cua and tRNA Cys/Cua. Proc.Natl.Acad.Sci. USA **83**, 6548-6552, 1986.

(4) Normanly, J., Kleina, L.G., Masson, J.M., Abelson, J. & Miller, J.H.: Construction of *Escherichia coli* amber suppressor tRNA genes. III Determination of tRNA specificity. J.Mol.Biol. **213**, 719-726, 1990.

(5) McClain, W.H. & Foss, K.: Changing the acceptor identity of a transfer RNA by altering nucleotides in a variable pocket. Science **241**, 1804-1807, 1988.

(6) Suckow, J., Markiewicz, P., Kleina, L.G., Miller, J., Kisters-Woike, B. & Müller-Hill, B.: Genetic studies of the Lac repressor XV: 4000 single amino acid exchanges. Analysis of the resulting phenotypes on the basis of the protein structure. J.Mol.Biol. in press, 1996.

(7) Miller, J.H. & Schmeissner, U.: Genetic studies of the *lac* repressor. X Analysis of missense mutations in the *lac I* gene. J.Mol.Biol. **131**, 223-248, 1979.

(8) Gordon, A.J.E., Burns, P.A., Fix, D.F., Yatagai, F., Allen, F.L., Horsfall, M.J., Halliday, J.A., Gray, J., Bernelot-Moens, C. & Glickman, B.W.: Missense mutation in the *lac I* gene of *Escherichia coli*. Inferences on the structure of the repressor protein. J.Mol.Biol. **200**, 239-251, 1988.

(9) Matthews, B.W., Ohlendorf, D.H., Anderson, W.F. & Takeda, Y.: Structure of the DNA-binding region of *lac* repressor inferred from its homology with *cro* repressor. Proc.Natl.Acad.Sci. USA **79**, 1428-1432, 1982.

(10) Kaptein, R., Zuiderweg, E., Scheek, R., Boelens. R. & van Gunsteren, W.: A protein structure from Nuclear Magnetic Resonance data, *lac* repressor headpiece. J.Mol.Biol. **182**, 179-182, 1985.

(11) Ebright, R.: Use of "loss-of-contact" substitutions to identify residues involved in an amino acid-base pair contact: Effect of substitutions of Gln 18 of *lac* repressor by Gly, Ser and Leu. J.Biomol.Struct.Dyn. **3**, 281-297, 1985.

(12) Ebright, R.: Evidence for a contact between glutamine-18 of *lac* repressor and base pair 7 of *lac* operator. Proc.Natl.Acad.Sci. USA **83**, 303-307, 1986.

(13) Boelens, R., Scheek, R.M., von Boom, J.H. & Kaptein, R.: Complex of *lac* repressor headpiece with a 14 base-pair *lac* operator fragment studied by two-dimensional nuclear magnetic resonance. J.Mol.Biol. **193**, 213-216, 1987.

(14) Lehming, N., Sartorius, J., Niemöller, M., Genenger, G., Wilcken-Bergmann, B.v. & Müller-Hill, B.: The interaction of the recognition helix of *lac* repressor with *lac* operator. EMBO J. **6**, 3145-3153, 1987.

(15) Lehming, N., Sartorius, J., Oehler, S., Wilcken-Bergmann, B.v.: Recognition helices of *lac* and λ *CI* repressor are oriented in opposite directions and recognize similar DNA sequences. Proc.Natl.Acad.Sci. USA **85**, 7947-7951, 1988.

(16) Kolkhof, P., Teichmann, D., Wilcken-Bergmann, B.v. & Müller-Hill, B.: Lac repressor with the helix-turn-helix motif of λ cro binds to lac operator. EMBO J. **11**, 3031-3038, 1992.

(17) Sartorius, J., Lehming, N., Kisters, B., Wilcken-Bergmann, B.v.: Lac repressor mutants with double or tripple exchanges in the recognition helix bind specifically to *lac* operator variants with multiple exchanges. EMBO J. **8**, 1265-1270, 1989.

(18) Lehming, N., Sartorius, J., Kisters-Woike, B., Wilcken-Bergmann, B.v. & Müller-Hill, B.: Mutant *lac* repressors with new specificities hint at rules for protein-DNA-recognition. EMBO J. **9**, 615-621, 1990.

(19) Sartorius, J., Lehming, N., Kisters-Woike, B., Wilcken-Bergmann, B.v. & Müller-Hill, B.: The roles of residues 5 and 9 of the recognition helix of Lac repressor in *lac* operator binding. J.Mol.Biol. **218**, 313-321, 1991.

(20) Müller-Hartmann, H. & Müller-Hill, B.: Side chains of the amino acid in position 110 of Lac repressor influence its allosteric equilibrium. J.Mol.Biol. **257**, 473-478, 1996.

(21) Schmitz, A., Schmeissner, U. & Miller, J.H.: Mutations affecting the quarternary struc-
      ture of the *lac* repressor. J.Biol.Chem. **251**, 3359-3366, 1976.
(22) Dong, F., Zimmermann, O. & Müller-Hill, B.: unpublished observations.
(23) Mandal, N., Su, W., Haber, R., Adhya, S. & Echols, H.: DNA looping in cellular repres-
      sion of transcription in the galactose operon. Genes and Development **4**, 410-418, 1990.
(24) Brenowitz, M., Mandal, N., Pickar, A., Jamison, E. & Adhya, S.: DNA-binding properties
      of a *lac* repressor mutant incapable of forming tetramers. J.Biol.Chem. **266**, 1281-1288,
      1991.
(25) Chakerian, A.E., Tesmer, V.M., Manly, S.P., Brackett, J.K., Lynch, M.J., Hoh, J.T. &
      Matthews, K.: Evidence for a leucine zipper motif in lactose repressor protein.
      J.Biol.Chem. **266**, 1371-1374, 1991.
(26) Alberti, S., Oehler, S., Wilcken-Bergmann, B.v., Krämer, H. & Müller-Hill, B.: Dimer-to-
      tetramer assembly of Lac repressor involves a leucine heptad repeat. The New Biologist
      3, 57-62, 1991.
(27) Alberti, S., Oehler, S., Wilcken-Bergmann, B.v.. & Müller-Hill, B.: Genetic analysis of
      the leucine heptad repeats of Lac repressor: evidence for a four helical bundle. EMBO
      J. **12**, 3227-3286, 1993.

3.4.4 RNA Polymerase, a Partner of Lac Repressor

When Jacob and Monod wrote their fundamental papers on repression (section
1.7.) they assumed the simplest possible mechanism: RNA polymerase and repressor compete with each other when they bind to their sites. Gilbert, his students and others demonstrated that *lac* promoter and operator indeed overlap
(section 1.18.). The center of *lac* operator lies 10.5 base pairs downstream from
the first transcribed base. Thus Lac repressor hinders RNA polymerase in binding
to its promoter.

Later two additional operators (*O2* and *O3*) were found . It took some time
for their function to be understood (see section 3.4.2.). If one sticks to the picture that *O1* and *P1* overlap and thus do not allow the simultaneous binding of
both RNA polymerase and Lac repressor, the picture does not change. The alleged roadblock at *O2* (1) plays only a very minor role in repression (2). The existence of an additional promoter *P2*, located 22 bp upstream of *P1*, complicates
the picture, but does not change it. In the presence of CAP protein *P2* is not occupied by RNA polymerase. When it is occupied by RNA polymerase very few
or no starts are observed (section 3.3.2.).

It has been claimed that Lac repressor is a transient gene activating protein (3).
I think the data are misinterpreted. It has also been claimed that RNA polymerase

binds to *P1* UV5 when Lac repressor is bound to *O1* (4). Again I think the data are misinterpreted. In each case the authors have not excluded RNA polymerase binding to *P2*.

A recent kinetic analysis of repression supports the assumption that Lac repressor acts by competing with RNA polymerase (5). Record's data exclude the more complex mechanisms. They support the simple mechanism of competition, which was also found in the case of the LexA (6) and λ *CI* repressor (7).

Is steric hindrance the only well documented mechanism of repression? I would like to include here the well documented case of GalR repression. In the *gal* system the two operators O_E and O_I lie too far upstream or downstream, respectively, from the promoter to compete directly with RNA polymerase. However, when a DNA loop is formed *in vitro* by the analogous tetrameric Lac repressor, the looped DNA becomes so rigid that it cannot bind a RNA polymerase molecule to its promoter site (8). It remains to be seen how this system, where no loop has been documented, works *in vivo*.

References

(1) Deuschle, U., Gentz, R. & Bujard, H.: *lac* repressor blocks transcribing RNA polymerase and terminates transcription. Proc.Natl.Acad.Sci. USA **83**, 4134-4137, 1986.

(2) Oehler, S., Eismann, E.R., Krämer, H. & Müller-Hill, B.: The three operators of the *lac* operon cooperate in repression. EMBO J. **9**, 973-979, 1990.

(3) Straney, S. & Crothers, D.M.: Lac repressor is a transient gene-activating protein. Cell **51**, 699-707, 1987.

(4) Lee, J. & Goldfarb, A.: *lac* repressor acts by modifying the initial transcribing complex so that it cannot leave the promoter. Cell **66**, 793-798, 1991.

(5) Schlax, P.J., Capp, M.W. & Record, T.M.: Inhibition of transcription initiation by *lac* repressor. J.Mol.Biol. **245**, 331-350, 1995.

(6) Bertrand-Burggraf, E., Hurstel, S., Daune, M. & Schnarr, M.: Promoter properties and negative regulation of the *uvrA* gene by the LexA repressor and its amino-terminal DNA binding domain. J.Mol.Biol. **193**, 293-302, 1987.

(7) Hawley, B.C., Johnson, A.D. & McClure, W.R.: Functional and physical characterisation of transcription initiation complexes in the bacteriophage λ O_R region. J.Biol.Chem. **260**, 8618-8626, 1985.

(8) Choy, H.E., Park, S.W., Parpack, P. & Adhya, S.: Transcription regulation by inflexibility of promoter DNA in a looped complex. Proc.Natl.Acad.Sci. USA **92**, 7327-7331, 1995.

3.5 β-Galactosidase, Lac Permease and Transacetylase

Only two of the three proteins of the *lac* operon have raised some real interest: β-galactosidase and Lac permease. The third, transacetylase, has remained in their shadow.

3.5.1 β-Galactosidase

For three reasons β-galactosidase captivated the interest of many biochemists: it is one of the largest proteins, it is easy to assay and the reaction it catalyses is simple to understand. To determine its protein sequence became a real challenge. At the time, in 1978, it was with its 1023 residues the largest protein sequence ever determined (1). It has remained the largest protein sequence ever determined, since DNA sequencing became available around this time. DNA sequencing is so much faster. Finally when the DNA sequence of the *lacZ* gene was determined, it demonstrated that the protein sequence had been almost correct. Only two minor mistakes were discovered (2).

To determine the X-ray structure of tetrameric β-galactosidase became the next challenge. The structure was solved in 1994 by Brian Matthews and his collaborators (3). Each subunit consists of the α peptide (see section 1.22.) and five domains. The α peptide which is capable of α complementation extends from residue 1 to 51. Domain one extends from residue 52 to 217, domain two from residue 220 to 334, domain three from residue 334 to 627, domain four from residue 627 to 736, and domain five from residue 737 to 1023. Topologically the second domain is identical with the fourth domain. In this structure all that is known about α complementation and N- or C-terminal fusions (see section 1.22.) makes sense (3).

An inspection of the structure confirms what genetic analysis had already suggested. It is impossible to delete any of its domains or large parts of any domain and still retain an active enzyme. So the question arises: why is β-galactosidase so large? After all, there are enzymes, *e.g.* some lysozymes which hydrolize glycosidic bonds and which are not larger than a hundred amino acid residues. An inspection of the protein sequences of various glycosidases shows that more than forty families exist (4). Some have been arranged into superfamilies but at the moment it is far from clear how many primary solutions of the glycosidase problem nature has produced .

At the moment it seems as though β-galactosidase of *E. coli* is a typical accident of evolution. This construct of five domains may have shown some low enzymatic activity and may have been improved by selection. Today it is impossible to look at the structure of β-galactosidase of *E. coli* and predict the structure of a much smaller variant showing similar enzymatic activity. We simply do not understand how this enzyme works and how proteins fold. This is a challenge for the future.

3.5.2 Lac Permease

Lac permease has been an interesting subject of research ever since its discovery (see section 1.4.). It transports various α or β galactosides into the *E. coli* cell against a concentration gradient. It does so by always co-transporting one proton together with one galactoside-molecule. The DNA of the *lacY* gene was sequenced in 1980 (5). The sequence suggested that Lac permease consists of 417 amino acid residues. Sequence and topological analyses suggest that Lac permease consists of twelve lipophilic α-helices (6). Both the N- and the C-termini lie on the inside of the *E. coli* cell. Extensive analysis indicated that Lac permease is active as monomer (7). Many residues, which are important or unimportant for activity, have been determined (8). The arrangement of the lipophilic α-helices is emerging (9).

Nonetheless, the structure and mode of function of this molecular machine are still unclear. Lac permease must exist in two extreme forms. In one form, the galactoside and the proton should be bound specifically to regions of Lac permease pointing toward the periplasmic space. In the other form, the galactoside and the proton should be bound specifically to regions of Lac permease pointing into the cytoplasma. Therefore to understand transport one needs the X-ray structure of both forms. Lac permease can now be easily purified in large amounts (9). So, obtaining the material is not the problem. One of the two forms can be stabilized by addition of a special galactoside which is bound, but not transported, by Lac permease (10). A major breakthrough is needed to prepare adequate crystals to solve the X-ray structures.

3.5.3 Thiogalactoside Transacetylase

There were always people on hand who were interested in β-galactosidase or Lac permease. But only very few have been seriously interested in transacetylase. It is an enzyme whose function is unclear. Does it serve to acetylate antibiotics (12)? Its sequence is vaguely similar to chloramphenicol acetyltransferase, type III, for which a three-dimensional structure is available (13,14). Both enzymes are homotrimers (15). After the entire genome of *Mycoplasma genitalium* was sequenced it came as a surprise that a gene weakly homologous to thiogalactoside transacetylase was found, but not the corresponding genes for β-galactosidase or Lac permease (16). Thiogalactoside transacetylase seems to fulfill a function in *Mycoplasma*, an organism which manages to replicate using only 470 genes. For *Mycoplasma* β-galactosidase and Lac permease seem to be unimportant. Yet it does not follow that the *lac* system may have evolved late after the mammals, the only living organisms producing lactose, had evolved. Plants are full of β-galactosido-glycerol derivates which also are substrates of the *lac* system (17). The history of the *lac* system is unknown.

References

(1) Fowler, A. & Zabin, I.: Amino acid sequence of β-galactosidase. J.Biol.Chem. **253**, 5521-5525, 1978.

(2) Kalnins, A., Otto, K., Rüther, U. & Müller-Hill, B.: Sequence of the *lacZ* gene of *E. coli*. EMBO J. **2**, 593-597, 1983.

(3) Jacobson, R.H., Zhang, X.J., DuBose, R.F. & Matthews, B.W.: Three-dimensional structure of β-galactosidase from *E. coli*. Nature **369**, 761-766, 1994.

(4) Henrissat, B. & Bairoch, A.: New families in the classification of glycosyl hydrolases based on amino acid sequence similarities. Biochem.J. **293**, 781-788, 1993.

(5) Büchel, D.E., Gronenborn, B. & Müller-Hill, B.: Sequence of the lactose permease gene. Nature **283**, 541-545, 1980.

(6) Calamia, J. & Manoil, C.: *Lac* permease of *Escherichia coli*: topology and sequence elements promoting membrane insertion. Proc.Natl.Acad.Sci. USA **87**, 4937-4941, 1990.

(7) Sahin-Tóth, M., Lawrence, M.C. & Kaback, R.: Properties of permease dimer, a fusion containing two lactose permease molecules from *Escherichia coli*. Proc.Natl.Acad.Sci. USA **91**, 5421-5425, 1994.

(8) Sahin-Tóth, M. & Kaback, R.: Cysteine scanning mutagenesis of putative transmembrane helices IX and X in the lactose permease of *Escherichia coli*. Protein Science **2**, 1024-1033, 1993.

(9) Wu, J., Perrin, D.M., Sigman, D.S. & Kaback, H.R.: Helix packing of lactose permease in *Escherichia coli* studied by site-directed chemical cleavage. Proc.Natl.Acad.Sci. USA **92**, 9186-9190, 1995.

(10) Consler, T.G., Person, B.L., Jung, H., Zen, K.H., Privé, G.G., Verner, G.E. & Kaback, R.: Properties and purification of an active biotinylated lactose permease from *Escherichia coli*. Proc.Natl.Acad.Sci. USA **90**, 6934-6938, 1993.

(11) Seibert, C., Dörner, W. & Jähnig, F.: A nontransportable substrate for lactose permease. Biochemistry **34**, 7819-7824, 1995.

(12) Andrews, K.J. & Lin, E.C.C.: Thiogalactoside transacetylase of the lactose operon as an enzyme for detoxification. J.Bact. **128**, 510-513, 1976.

(13) Leslie, A.G.W., Moody, P.C.E. & Shaw, W.V.: Structure of type III chloramphenicol acetyltransferase at 1.75-Å resolution. Proc.Natl.Acad.Sci. USA **85**, 4133-4137, 1988.

(14) Leslie, A.G.W.: Refined crystal structure of type III chloramphenicol acetyltransferase at 1.75 Å resolution. J.Mol.Bol. **213**, 167-186, 1990.

(15) Lewendon, A., Ellis, J. & Shaw, W.V.: Structural and mechanistic studies of galactosidase acetyltransferase, the *Escherichia coli LacA* gene product. J.Biol.Chem. **270**, 26326-26331, 1995.

(16) Fraser, C.M., Gocayne, J.D., White, O., Adams, M.D., Clayton, R.A., Fleischmann, R.D., Bult, C.J., Kerlavage, A.R., Sutton, G., Kelley, J.M., Fritschman, J.L., Weidman, J.F., Small, K.V., Sandusky, M., Fuhrmann, J., Nguyen, D., Utterback, T.R., Saudeck, D.M., Philipps, C.A., Merrick, J.M., Tomb, J.F., Dougherty, B.A., Bott, K.F., Hu, P.C., Lucier, T.S., Peterson, S.N., Smith, H.O., Hutchison III, C.A. & Venter, J.C.: The minimal gene complement of *Mycoplasma genitalium*. Science **270**, 397-403, 1995.

(17) Müller-Hill, B.: Lac repressor. Angewandte Chemie int.ed. **10**, 160-172, 1971.

3.6 The *lac* System as a Tool

Five major practical uses have been made of parts of the *lac* system: 1. The *lacZ* gene coding for β-galactosidase has been fused to many other genes. 2. α complementation has been used for cloning. 3. The *lacI* gene has been used as an indicator of mutagenesis. 4. The *lacI/O1* control system has been used in other prokaryotes or in eukaryotes. 5. The *lacI/O1* system has been used for peptide libraries.

3.6.1 β-Galactosidase Fusions and Lac Repressor Fusions

The $I^{Q}Z^{-u118}$ double mutant carries a mutation in the promoter of the *I* gene which leads to a tenfold overproduction of Lac repressor. In addition it carries an ochre mutation at codon 17 of *lacZ* coding for β-galactosidase. If one plates such bacteria on lactose plates, they do not grow. However one observes occasional *lac⁺* revertants, some of which are constitutive. These have fusions joining the *I* gene with the *Z* gene. This is possible since the N-terminal 23 residues of β-galactosidase are not necessary for enzymatic activity and the C-terminal 30 residues of Lac repressor are not necessary for the formation of a functional Lac repressor dimer. Thus, since the architecture of the C-terminus of Lac repressor and of the N-terminus of β-galactosidase allow it, one can isolate revertants which have a functional, dimeric Lac repressor fused to a functional β-galactosidase (1). If one uses a *lac* Z-containing transposon such *in vivo* fusions can also be obtained with many other *E. coli* genes (2). But, of course, the fusions are easier to construct *in vitro*, following the introduction of restriction sites in the extreme 5' end of the *lacZ* gene (3). Restriction sites allowing active fusions have also been introduced at the extreme 3' end of the *lacZ* gene (4).

The commercial availability of a soluble colourless substrate of β-galactosidase which is hydrolysed to an insoluble coloured (blue) product (5-bromo-4-chloro-3-indolyl-β-D-galactoside) has made β-galactosidase one of the most commonly used *in vivo* reporter enzymes. *In vitro* one easily can test β-galactosidase activity with a colourless substrate which is hydrolysed to a yellow soluble product (o-nitrophenyl-β-D-galactoside). So one can use the *lacZ* gene as a reporter of the regulation of any promoter. One can fuse the *lacZ* gene with a minimal eukaryotic promoter on a transposable element in *Drosophila* and isolate strains on which this minimal promoter has been inserted close to the regulatory elements of an existing promoter. β-galactosidase will then be produced under the control of the elements of this promoter. However it will remain in the cytoplasm. One can fuse the *lacZ* gene with elements which code for peptide domains which direct transport. Thus, for example the *lacZ* gene has been fused with DNA coding for Tau which targets the β-galactosidase to axons (5). It is here not appropriate to list all the cases where β-galactosidase has been used as a reporter. It suffices to open any recent issue of any journal dealing with eukaryotic molecular biology to get an impression. Finally, Lac repressor fusions can be used to direct the fused enzyme to become active close to a *lac* operator. This has been demonstrated with phage T7 endonuclease which was fused to the extreme 3' end of Lac repressor (6).

3.6.2 α-Complementation

A *lacZ* gene which lacks codons 11 to 41 produces completely inactive, dimeric β-galactosidase. Active tetramer is restored when a peptide consisting of the N-terminal 60 residues of β-galactosidase is added either *in vitro* or *in vivo*. This phenomenon, called α-complementation (see section 1.22.), has been used successfully to indicate cloning. For example a short DNA fragment consisting of the promoter region and the first sixty codons of the *lacZ* gene has been cloned into the DNA of phage *M13* (7). This phage now produces active α peptide which can be used for α-complementation. If one clones foreign DNA somewhere into the DNA coding for the α peptide one will destroy the β-galactosidase activity. Agnes Ullmann who discovered α-complementation has written an instructive essay on the history and present day use of the phenomenon (8).

3.6.3 The *lacI* Gene as an Indicator of Mutagenesis

Mutations in the *lacI* gene are so easy to detect, to isolate and thus to sequence that the *I* gene was and is used extensively to study the process of mutagenesis (9,10). Accordingly the mutations generated by the SOS system (11), by the *mutY* gene (12), the *mutA* and the *mutC* genes (13) were so determined. The mutagenic action of various carcinogens in *E. coli* was similarly studied (14,15). Ultimately *lacI* was introduced on suitable vectors into human cells, mutagenized and retransferred into *E. coli* and sequenced (16,17). So far the best, *i.e.* the fastest of such techniques makes use of PCR-single strand conformation polymorphism analysis (18).

3.6.4 *Lac* Control not in *E. coli*

The *lac* system is very efficiently controlled in *E. coli*. Thus β-galactosidase can be induced one thousand fold in the wild type situation. When *lac* operator *O1* is placed downstream of a promoter of *B.subtilis*, a hundred fold repression was observed when Lac repressor was present in large amounts (19). When a *lac* operator is properly placed within a eukaryotic animal (20) or plant (21) promoter, repression is ten to a hundred fold, depending on the amount of Lac repressor produced in these cells. In these experiments with eukaryotic systems Lac repressor lacks a signal which directs it into the nucleus. Such a signal has been fused to the N-terminus of Lac repressor without disturbing its capacity to bind to *lac* operator (22). But repression still does not reach the level observed in *E. coli*. The reason for this is clear. The eukaryotic systems lack properly positioned main and properly spaced auxiliary operators.

What is required is a eukaryotic promoter which is at will completely, *i.e.* a thousand fold repressed. Two approaches have been tried. The transcription of the gene in question was made dependent on coliphage T3 RNA polymerase, and the production of the T3 RNA polymerase was put under the control of Lac repressor (23). This type of construct was recently optimized (24). In this optimized construct the reporter enzyme β-galactosidase can be induced 10,000 fold. In a second attempt, an activating domain was added to the extreme C-terminus of Lac repressor. This destroys the signal for tetramerisation, but still produces intact Lac repressor dimers (25). Such a construct has been used to select homozygous negative mutants in mouse cells and so to detect a novel tumor susceptibility gene (26). The best simple control system for a eukaryotic cell is currently a similar construct using Tet repressor (27). It remains to be seen whether a similarly effective design of a Lac control system is feasible.

3.6.5 Peptide Libraries Displayed on Dimeric Lac Repressor

The peptide libraries obtained with filamentous phage are effective because each peptide is physically connected with the phage DNA which codes for it. Similarly if one presents the peptide at the extreme C-terminus of Lac repressor (28) a peptide library can be made to stick to the plasmid DNA which encodes for it. Lac repressor remains an active dimer and sticks sufficiently well to the particular ideal *lac* operator containing DNA which encodes for it. Under appropriate conditions the repressor-operator bond is tight enough so that it does not open when the bacteria are lysed. This method has been refined (29) and used practically (30). It remains to be seen how large such libraries may become.

References

(1) Müller-Hill, B. & Kania, J.: Lac repressor can be fused to β-galactosidase. Nature **249**, 561-563, 1974.

(2) Silhavy, T.J., Berman, M.L. & Enquist, L.M. (Eds.): Experiments with gene fusions. Cold Spring Harbor Laboratory Press, 1984.

(3) Messing, J. & Gronenborn, B.: Methylation of single-stranded DNA *in vitro* introduces new restriction endonuclease cleavage sites. Nature **272**, 375-377, 1978.

(4) Rüther, U. & Müller-Hill, B.: Easy identification of cDNA clones. EMBO J. **10**, 1791-1794, 1983.

(5) Callahan, C.A. & Thomas, J.B.: Tau-β-galactosidase, an axon-directed fusion protein. Proc.Natl.Acad.Sci. USA **91**, 5972-5976, 1994.

(6) Panayotatos, N., Fontaine, A. & Bäckman, S.: Biosynthesis of a repressor/nuclease hybride protein. J.Biol.Chem. **264**, 15066-15069, 1989.

(7) Messing, J., Gronenborn, B., Müller-Hill, B. & Hofschneider, P.H.: Filamentous coliphage *M13* as a cloning vehicle: insertions of a HindII fragment of the *lac* regulatory region in *M13* replicative form *in vitro*. Proc.Natl.Acad.Sci. USA **74**, 3542-3646, 1977.

(8) Ullmann, A.: Complementation in β-galactosidase: from protein structure to genetic engineering. Bioessays **14**, 201-205, 1992.

(9) Albertine, A.M., Hofer, M., Calos, M.P. & Miller, J.H.: On the formation of spontaneous deletions: the importance of short sequence homologies in generation of large deletions. Cell **29**, 319-326, 1982.

(10) Tlsty, T.D., Albertini, A.M. & Miller, J.H.: Gene amplification on the *lac* region of *E. coli*. Cell **37**, 217-224, 1984.

(11) Miller, J.H. & Low, K.B.: Specificity of mutagenesis resulting from the induction of the SOS system in the absence of mutagenetic treatment. Cell **37**, 675-682, 1984.

(12) Nghiem, Y, Cabrera, M., Cupples, C.G. & Miller, J.H.: The *mutY* gene: A mutator locus in *Escherichia coli* that generates G.C -T.A transversions. Proc.Natl.Acad.Sci. USA **85**, 2709-2713, 1988.

(13) Michaelis, M.L., Cruz, C. & Miller, J.H.: *mutA* and *mutC*: Two mutator loci in *Escherichia coli* that stimulate transversions. Proc.Natl.Acad.Sci. USA **87**, 9211-9215, 1990.

(14) Eisenstadt, E., Warren, J.A., Porter, J., Atkins, D. & Miller, J.H.: Carcinogenic epoxides of benzo(*a*)pyrene and cyclopenta(*cd*)pyrene induce base substitutions via specific transversions. Proc.Natl.Acad.Sci. USA **79**, 1945-1949, 1982.

(15) Miller, J.H.: Carcinogens induce targeted mutations in *Escherichia coli*. Cell **31**, 5-7, 1982.

(16) Lebkowski, J.S., Clancy, S., Miller, J.H. & Calos, M.: The *lacI* shuttle: Rapid analysis of the mutagenic specificity of ultraviolet light in human cells. Proc.Natl.Acad.Sci. USA **82**, 8606-8610, 1985.

(17) Hsia, H.C., Lebkowski, J.S., Leong, P.M., Calos, M.P. & Miller. J.H.: Comparison of ultraviolet irridation-induced mutagenesis of the *lacI* gene in *Escherichia coli* and in human 293 cells. J.Mol.Biol. **205**, 103-113, 1989.

(18) Ushijima, T., Hosoya, Y., Suzuki, T., Sofuni, T., Sugimura, T. & Nagao, M.: A rapid method for detection of mutations in the *lacI* gene using PCR-single strand conformation polymorphism analysis: demonstration of its high sensitivity. Mutation Res. **334**, 283-292, 1995.

(19) Yansura, D.G. & Henner. D.J.: Use of the *Escherichia coli lac* repressor and operator to control gene expression in *Bacillus subtilis*. Proc.Natl.Acad.Sci. USA **81**, 439-443, 1984.

(20) Brown, M., Figge, J., Hansen, U., Wright, C., Jeang, K.T., Khoury, G., Livingston, D.M. & Roberts, T.M.: *lac* repressor can regulate expression from a hybrid SV40 early promoter containing *lac* operator in animal cells. Cell **49**, 603-612, 1987.

(21) Wilde, R.J., Shufflebottom, D., Cooke, S., Jasinka, I., Merryweather, A., Beri, R., Brammer, W.J., Bevan, M. & Schuch, W.: Control of gene expression in tobacco cells using a bacterial operator-repressor system. EMBO J., **11**, 1251-1259, 1992.

(22) Fieck, A., Wyborski, D.L. & Short, J.M.: Modification of the *E. coli* Lac repressor for expression in eucaryotic cells: effects of nuclear signal sequences on protein activity and nuclear accumulation. Nucl.Acids Res. **20**, 1785-1791, 1992.

(23) Deuschle, U., Pepperkok, R. Wang, F., Giordano, T.J., McAllister, W.T., Ansorge, W. & Bujard, H.: Regulated expression of foreign genes in mammalian cells under the control of coliphage T3 RNA polymerase and *lac* repressor. Proc.Natl.Acad.Sci. USA **86**, 5400-5404, 1989.

(24) Ward, G.A., Stover, C.K., Moss, B., Fuerst, T.R.: Stringent chemical and thermal regulation of recombinant gene expression by vaccinia virus vectors in mammalian cells. Proc.Natl.Acad.Sci. USA **92**, 6773-6777,1995.

(25) Baim, S.B., Labow, M.A., Levine, A.J. & Shenk, T.: A chimeric mammalian transactivator based upon the *lac* repressor that is regulated by temperature and isopropyl-β-D-thiogalactoside. Proc.Natl.Acad.Sci. USA **88**, 5072-5076, 1991.

(26) Li, L. & Cohen, S.N.: tsg101: a novel tumor susceptibility gene isolated by controlled homozygous functional knockout of allelic loci in mammalian cells. Cell **85**, 319-329, 1996.

(27) Gossen, M., Freundlieb, S., Bender, G., Müller, G., Hillen, W. & Bujard, H.: Transcription activation by tetracyclines in Mammalian cells. Science **268**, 1766-1769, 1995.

(28) Cull, M.G., Miller, J.F. & Schatz, P.J.: Screening for receptor ligands using large libraries of peptides linked to the C-terminus of the *lac* repressor. Proc.Natl.Acad.Sci. USA **89**, 1865-1869, 1992.

(29) Gates, C.M., Stemmer, W.P.C., Kaptein, R. & Schatz, P.J.: Affinity selective isolation of ligands from peptide libraries through display on *lac* repressor "headpiece dimer". J. Mol. Biol. **255**, 373-386, 1996.

(30) Martens, C.L., Cwirla, S.E., Lee, R.Y.W., Whitehorn, E., Chen,, E.Y.F., Bakker, A., Martin, E.L., Wagstrom, C., Gopalan, P. Smith, C.W., Tate, E., Koller, K.J., Schatz, P.J., Dower, W.J. & Barret, R.W.: Peptides which bind to E-selectin and block neutrophil adhesion. J.Biol.Chem. **270**, 21129-21136, 1995.

3.7 What Lesson can be Learned from the *lac* System?

Having read the previous chapters on the *lac* system one may get an uneasy feeling that all is beautiful and interesting but rather special and therefore irrelevant to a real understanding of gene expression. There is a good argument against this notion. The proteins of the *lac* system are not unique, but members of large families. Let's document this with the best analysed protein, Lac repressor. A recent paper (1) indicates that to date, nine other *E. coli* repressors have been found which are similar in sequence: two Gal repressors, the Cyt repressor, the Pur repressor, the Ebg repressor, the Raf repressor, the Fru repressor, the Mal repressor and the Asc repressor. This is not all.

E. coli produces three more proteins, the galactose, ribose and the arabinose binding proteins, which are similar in sequence to that part of Lac repressor coding for the core, the domain which binds inducer. There are two more proteins in *E. coli* which have no sequence similarity, but which have the same three-dimensional structure (see section 3.4.1.). The *AmiC* protein, a controller of transcription anti-termination in *Pseudomonas aruginose* was shown by X ray analysis to have the same structure (2). There are at least three eukaryotic proteins which have the same structure too (3-5). A philosopher may say that the idea of the three dimensional structure is always the same but that the languages (protein sequences) in which it is expressed differ.

Moreover, the HTH motif of the headpiece of Lac repressor which binds to DNA is similar to the HTH motifs found in several phage repressors. In the phage repressors the HTH motifs carry an additional helix at their N-terminus and β sheets at their C-termini; and the headpieces of Lac and phage λ *CI* repressor differ in their assembly as dimers.

It is not clear whether this large family of proteins has been found only because of the predominant interest in several of its members. So it is not clear whether most proteins of *E. coli* are part of similar large families. A first analysis seems to contradict that (6). But still I would not be astonished if this were a general property of most proteins of *E. coli*, of all bacteria and similarly of most proteins of mouse and man.

Let us assume for a moment that all proteins of *E. coli* are members of similar large families. *E. coli* has 4.7×10^6 base pairs. This corresponds to a coding capacity of about $1.5 \ 10^6$ amino acids. Thus, about 5,000 different proteins of about MW 30,000 Daltons can be constructed. If all proteins belonged to families as large as the Lac repressor family, this would reduce the number of possible protein structures of *E. coli* to about 300. This is a number one can deal with. Three hundred structures with different functions where each function has a dozen or so possible variants can be understood and memorized.

Now we have to ask whether the other three proteins of the *lac* system, β-galactosidase, Lac permease and transacetylase, belong to equally large families. There exists only one other protein (Ebg β-galactosidase) which is fully homologous to Lac β-galactosidase. But we have to recall that Lac β-galactosidase is unusually large. Its 1,023 residues form five independent domains. One has to look for proteins homologous to one or the other of these domains. This analysis has not been done. I simply note here that uidA β-glucuronidase (7) is similar in sequence to the first three domains of β-galactosidase.

RafB permease (8), sucrose permease (9) and an unknown protein (10) are similar in sequence to Lac permease. These numbers are smaller than the corresponding numbers of Lac repressor. But the homology tests used were much less sensitive than the tests used in connection with Lac repressor. It is significant that the test did not reveal melibiose permease. Mel permease can be shown to contain twelve lipophilic membrane-crossing alpha helices like Lac permease (10). There are other permeases which share this property. Are they genuine relatives of Lac permease? I think so. Four proteins can be found which are rather similar in sequence to transacetylase.

If one looks at some of the systems regulated by related repressors and consisting of proteins catalysing similar reactions, one gets an interesting picture. I show here five systems: the *lac*, the *gal*, the *mel*, the *raf* and the *ara* systems (Fig. 20). If one compares the *lac* (β-gal) with the *mel* (α-gal) system one notices that the α-gal system is regulated by MelR, a relative of AraC, and a protein which is capable of both repression and activation. Three enzymes of the *gal* operon perform similar catalytic functions as do enzymes of the *araBAD* operon. Are the enzymes with similar functions similar in structure? We do not know. One may argue that their organisation is different in the two operons and that that speaks against a close relationship. But the *lac* operon is organized differently in one of the few other bacterial system which has been analysed. In *Klebsiella* the promoters of the *lacI* and the *lacZ* genes diverge (12,13)! If one compares the protein sequences of the *gal* and *ara* enzymes one observes a marginal similarity – which I think is significant.

Thus I propose that eventually it will be found that Gal epimerase will be similar in structure to ribulose-5-phosphate-epimerase (the epimerase of the *ara* system). I propose the same for the isomerases and kinases of these two systems. Should this born out, one may come to the conclusion that by understanding the

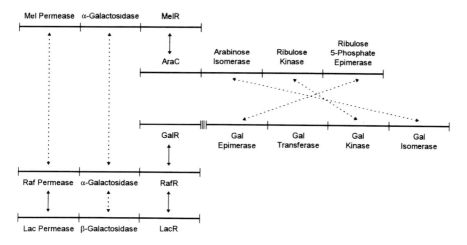

Fig. 20.: Four operons of *E.coli*. Similarity in structure: solid lines. Possible similarity in structure: broken lines.

lac system one understands about one percent of the proteins of *E. coli*. Is this a
lot or a little? I think it is a lot. Finally let me state that I am well aware that my
argument is not altogether new (14).

References

(1) Markiewicz, P., Kleina, L.G., Cruz, C., Ehret, S. & Miller, J.H.: XIV. Analysis of 4000 altered *Escherichia coli lac* repressors reveals essential and non-essential residues, as well
as "spacers" which do not require a specific sequence. J.Mol.Biol. **240**, 421-433, 1994.

(2) Pearl, L., O'Hara, B., Drew, R. & Wilson, C.: Crystal structure of AmiC: the controller of
transcription antitermination in the amidase operon of *Pseudomonas aeruginosa*. EMBO
J. **13**, 5810-5817, 1994.

(3) Anderson, B.F., Baker, H.M., Norris, G.E., Ruball, S.V. & Baker, E.N.: Apolactoferrin
structure demonstrates ligand-induced conformational change in tranferrins. Nature **344**,
784-787, 1990.

(4) O'Hara, P.J., Sheppard, P.O., Thogersen, H., Venezia, D., Haldeman, B.A., McGrane, V.,
Houamed, K.M., Thomsen, C., Gilbert, T.H. & Mulvihill, E.R.: The ligand-binding do-

main of metabotropic glutamate receptors is related to bacterial periplasmic binding protein. Neuron **11**, 41-52, 1993.

(5) Brown, E.M., Gamba, G., Riccardi, D., Lombardi, M., Butters, R., Kifor, O., Sun, A., Hediger, M.A., Lytton, J. & Hebert, S.: Cloning and characterization of an extracellular Ca^{2+}-sensing receptor from bovine parathyroid. Nature **366**, 575-580, 1993.

(6) Koonin, E.V., Tatusov, R.L. & Rudd, K.E.: Sequence similarity analysis of *Escherichia coli* proteins: functional and evolutionary implications. Proc.Natl.Acad.Sci. USA **92**, 11921-11925, 1995.

(7) Jefferson, R.A., Burgess, S.M. & Hirsch, D.: β-glucuronidase from *Escherichia coli* as a gene-fusion marker. Proc.Natl.Acad.Sci. USA **83**, 8447-8451, 1986.

(8) Aslanidis, C., Schmid, K. & Schmitt, R.: Nucleotide sequences and operon structure of plasmid-borne genes mediating uptake and utilization of raffinose in *Escherichia coli*. J.Bact. **171**, 6753-6763, 1989.

(9) Sahin-Toth, Frillingos, S., Lengeler, J.W. & Kaback, H.R.: Active transport by the *Csc* permease in *Escherichia coli*. Biochem.Biophys.Res.Commun. **208**, 1116-1123, 1995.

(10) Ueguchi, C. & Ito, K.: Multicopy suppression: an approach to understanding intracellular functioning of the protein export system. J.Bact. **174**, 1454-1461, 1992.

(11) Yazu, H., Shiota-Niiya, S., Shimamoto, T., Kanazawa, H., Futai, M. & Tsuchiya, T.: Nucleotide sequence of the *mel B* gene and characteristics of deduced amino acid sequence of the melibiose carrier in *Escherichia coli*. J.Biol.Chem. **259**, 4320-4326, 1984.

(12) Buvinger, W.E. & Riley, M.: Nucleotide sequence of *Klebsiella pneumoniae lac* genes. J.Bact. **163**, 850-857, 1985.

(13) Buvinger, W.E. & Riley, M.: Regulatory region of the divergent *Klebsiella pneumoniae lac* operon. J.Bact. **163**, 858-862, 1985.

(14) Chotia, C.: Proteins: one thousand families for the molecular biologist. Nature **357**, 543-544, 1992.

3.8 Outlook

The Pessimist's View

These days very few papers are published in *Cell* which deal with the *lac* system. And when a paper is published, it has to present an unconventional point of view. The latest in this line is by Diane M. Retallack and David I. Friedman (1). Its title summarizes its content: "A role for a small stable RNA in modulating the activity of DNA-binding proteins". The authors describe a RNA, the normal product of the *ssrA* gene, which *weakens* the Lac repressor (and other repressors) by a factor of about ten. In the absence of this RNA, the repressed level of β-galactosidase decreases to one tenth of the wild type level. In other words, repression increases by a factor of ten when the *ssrA* gene is knocked out. This result seems most interesting. However, when one actually looks at the numbers one finds that the *lac* system the authors studied is only repressed thirty to fifty fold in the *ssrA* wild type situation! In the absence of the *ssrA* RNA, the repression value increases to about three hundred or five hundred fold. Here in astonishment the knowledgeable reader will stop reading. By now he realizes that it is well known and documented in many, many papers that in the *lac* system the wild type repression factor is about one thousand! Thus the *lac ssrA* wild type the authors claim to have analysed was not the *lac* wild type.

  The authors then proceed to gel retardation experiments. Here they claim that they retarded 2 pmol labelled *ssrA* RNA with 7 pmol tetrameric Lac repressor. This looks fine, but now comes the surprise. When they add 20 pmol of cold competitor RNA, the shifted band completely disappears. Instead, a nonshifted band of about half the intensity of the band without Lac repressor is seen. This is contrary to quantitative expectation. If they had really used 7 pmol tetrameric Lac repressor, this would equal 14 pmol of Lac repressor dimers, which is presumably the unit which will interact with the RNA. In this case, the addition of 20 pmol cold RNA should only lead to a two fold decrease of the intensity of the shifted band. So something is wrong. Whoever reads the entire paper will find the rest equally dubious and incomprehensible. Is this good science? No, I do not think so. Who were the referees? Who was the editor? It is not reassuring that the authors thank about a dozen well known geneticists for discussion or helpful suggestions in writing the manuscript. The pessimist sees confusion growing and reason vanishing.

The Optimist's View

The optimist finds confusion funny. He laughs when he reads in the valediction of John Maddox (2): "Within a few months of Nature's centenary in 1969, for example, we had Mark Ptashne's discovery of the *lac* operon (the first genetic regulatory element to be recognized) …". The optimist does not care about the past. He asks the question: What will the future of the *lac* system of *E. coli* bring? The structural and functional work has not come to its end. I would like to see the X-ray structures of the CAP-RNA polymerase promoter complex and of Lac permease. I would like to understand how they function mechanically. So, I imagine in thirty years movies can be looked at on computer screens. There we will see how lactose molecules tumble around *E. coli* and finally enter through pores into the periplasmic space of an *E. coli* cell. We will see how they pass by the ribose and arabinose binding proteins. How one of them hits upon a Lac permease molecule and is drawn into the interior of the *E. coli* cell. How it there finally meets a β-galactosidase molecule. How it is almost hydrolysed but at the last moment transformed into allolactose. How the allolactose molecule escapes the β-galactosidase. How it finds an inducer binding site on a subunit of tetrameric Lac repressor bound to the main *lac* operator and one of the auxiliary *lac* operators. How the repressor molecule changes its shape and falls off the one operator. How CAP helps RNA polymerase to bind to the *lac* promoter. How RNA polymerase gets into action. How more RNA polymerase molecules bind and start. How the *lac* DNA is transcribed into RNA and how the RNA is translated first into β-galactosidase, then into Lac permease and finally into transacetylase. How more and more lactose molecules are drawn into the *E. coli* cell to be hydrolyzed into galactose and glucose. How the galactose interacts with dimeric Gal repressor bound to one of the two *gal* operators. How the *gal* genes are first transcribed and then translated. How the four enzymes of the *gal* operon catalyse the transformation of galactose into glucose. And there the film will end and everybody will be invited to see the next films on the life of phage Lambda etc.

The structures shown in the films will be accurate but of course coloured. Some of their possible variants will also be shown. One will learn how the various repressors recognize themselves and their DNA operators. I imagine that a time may come in a hundred years where it seems that all possible interesting work has been done with the *lac* system. All that will be left for students is to look at these historic TV movies. Will this take fifty or a hundred years? I do not know, but I am sure it will happen.

The *lac* system consists only of four different protein structures. More time will be needed until all the three hundred procaryotic and perhaps one thousand eukaryotic structures are solved. The various protein families will have to be allocated. And then the real work begins: to construct the complete network of all their interactions. Here, two poems by Goethe come to mind: The first one (3) is a warning:

Dummes Zeug kann man viel reden,
Kann es auch schreiben,
Wird weder Leib noch Seele tödten,
Es wird alles beim Alten bleiben.
Dummes aber vors Auge gestellt,
Hat ein magisches Recht;
Weil es die Sinne gefesselt hält,
Bleibt der Geist ein Knecht.

The second one (4) is an encouragement. I quote just five lines:

Und war es endlich dir gelungen,
Und bist du vom Gefühl durchdrungen:
Was fruchtbar ist allein ist wahr;
Du prüfst das allgemeine Walten,
Es wird nach seiner Weise schalten,
Geselle dich zur kleinsten Schar.

References

(1) Retallack, D.M. & Friedman, D.I.: A role for a small stable RNA in modulating the activity of DNA-binding proteins. Cell **83**, 227-235, 1995.
(2) Maddox, J.: Valediction from an old hand. Nature **378**, 521-523, 1995.
(3) Goethes Werke, Vollständige Ausgabe letzter Hand **47**, 73-74, 1833.
    One can say many stupid things/ one may also write them down/ it will never kill body or mind/ things will stay as they are./ But stupid things put in front of the eyes, have a magic right;/ they keep the senses chained/ the mind remains a slave.
(4) ibid. **3**, 268, 1828.
    And did you finally succeed,/ And were you filled by the feeling/ only what is fertile is true/ you look at the world/ and it will act as it acts/ now join the smallest group.

Authors Index

Subject Index